Organic Aquaculture

Giuseppe Lembo • Elena Mente
Editors

Organic Aquaculture

Impacts and Future Developments

Springer

Editors
Giuseppe Lembo
Stazione Sperimentale per lo Studio delle Risorse del Mare
COISPA Tecnologia & Ricerca
Bari, Italy

Elena Mente
School of Agricultural Sciences
University of Thessaly
Volos, Greece

ISBN 978-3-030-05602-5 ISBN 978-3-030-05603-2 (eBook)
https://doi.org/10.1007/978-3-030-05603-2

Library of Congress Control Number: 2019933898

This Springer imprint is published by the registered company Springer Nature Switzerland AG.
The registered company address is: Gewerbestrasse 11, 6330 Cham, Switzerland

Dedication of the first author
To my beloved Nuccia

Dedication of the second author
To my nephew Konstantinos

Preface

> Let food be the medicine and medicine be the food. (Hippocrates of Kos (460 B.C.–377 B.C.))

In 2015, world aquaculture production reached 106 million tonnes, while European aquaculture production reached 2.98 million tonnes of seafood, with a value over 11 billion USD (FAO, 2017). Although the world aquaculture production is still the fastest-growing food-producing sector in the world, the European one increased by about 136% in the last 10 years. Organic aquaculture represents 4.7% of the total EU aquaculture production, with six main species: salmon, trout, carp, sea bass, sea bream and mussels. In addition, organic aquaculture has experienced a significant increase in recent years. In 2015, compared to 2012, it grew by 24% for salmon, doubled for rainbow trout and tripled for sea bass and sea bream (EUMOFA, 2017). Positive increase has also been observed for shellfish (mussels, oysters).

The main species produced under organic standards in 2015 were salmon, more than 16.000 tonnes (9% of EU total salmon production), the main producers being Ireland and the UK; mussel, almost 20.000 tonnes (4% of EU total mussel production), Ireland, Italy and Denmark being the main producers; carp, about 6.000 tonnes (8% of EU total carp production), the main producers being Hungary, Romania and Lithuania; trout, almost 5.000 tonnes (3% of EU total trout production), the main producers being France and Denmark; and sea bass and sea bream, about 2.000 tonnes (1% of EU total production), the main producers being France, Greece and Spain.

Yet, the economic performance of EU organic aquaculture seems far from being satisfactory everywhere (EUMOFA, 2017). Notwithstanding the above, Europe is still heavily dependent on external markets to cover seafood demand. Indeed, EU seafood imports account to 148% of the EU production. Thus, EU aquaculture needs to increase its products' competitiveness and respond to consumer demands for high-quality and safe food, in a challenging context of climate change, greater competition for natural resources and conflicting interests for space and markets. To ensure food and nutrition security by 2020, the food production sectors have to sustainably expand in terms of space use, production and new value chains, exploring and enhancing innovation opportunities offered by sustainable and resilient aquaculture production systems, implementing the circular economy principles and increasing social acceptance of the corresponding activities and products. Future aquaculture needs to manage production of high-quality, safe food for a growing population, without harming the environment and without compromising animal health and welfare. To succeed, improvements are demanded along the whole value chain, including a sustainable, organic, resilient and friendly aquaculture.

Organic aquaculture is an alternative production approach driven by the growing interest in sustainable utilization of resources. It is a system of farm management and food production that combines best environmental practices, a high level of biodiversity, the preservation of natural resources, the application of high animal welfare standards and a production method in line with the preference of certain consumers for products produced using natural substances and processes. Thus, it plays a dual societal role, where on the one hand provides for a specific market responding to a consumer demand for high-quality organic products and on the other hand delivers public goods contributing to the protection of the environment and animal welfare, as well as to rural development (EC 834/2007).

Organic-certified aquaculture products include fish, molluscs, crustaceans and algae (seaweed and phytoplankton). This book addresses, reviews and evaluates key themes and is set out to show how these themes relate to the challenges and bottleneck for a sustainable organic aquaculture production in Europe. The key themes reflect the main challenges facing the organic aquaculture industry: guarantee and certification system, nutrition, reproduction, production system design and animal welfare. In addition, it assesses the impact of new and future potential development of new knowledge to update and modify the criteria and standards for organic aquaculture to provide high-quality products for the consumers.

The overview of the book is followed by eight chapters.

Chapter 1 provides an overview of the organic standards and certification systems worldwide and presents the organic principles. Organic systems are rooted in four principles of organic agriculture, "*health, ecology, fairness* and *care*". The goal of the organic sector is to positively impact aquaculture production through the establishment of ecologically integrated systems that preserve the natural environment, maintain or enhance biodiversity, respect animal welfare and yield high-quality, healthy product. Organic standards and regulations, in order to guarantee and protect consumers, are based on a guarantee or assurance system, which focuses on assuring certain basic facets of the standards that have been met for the relevant products in the marketplace. The implication is that the processes (practices) followed by the operators along the chain lead to products with a degree of certainty that they possess specific attributes, e.g. have been produced without deliberate use of genetic engineering, harmful pesticides and hormones and have respected aspects of animal welfare, biodiversity and, in the case of certain private standards in particular, social justice issues. Furthermore, most certification systems (e.g. EU regulation) require traceability from final product back through the chain of custody.

Chapter 2 deals with the European Regulation on organic aquaculture and goes into the details of the rules, with reference to (a) the origin of the aquaculture animals, (b) the husbandry practices, (c) the breeding, (d) the feed for fish and crustaceans, (e) the bivalve molluscs and other species which are not fed by man but feed on natural plankton, (f) the disease prevention and veterinary treatment and (g) the cleaning and disinfection, products for cleaning and disinfection in ponds, cages, buildings and installations. In addition to the official controls along the agri-food chain applied to both organic and non-organic food and feed marketed, the EU organic aquaculture production, trade and labelling is subject to an additional control system established in the Regulation (EC) No. 834/2007 and its implementing regulations. Such control system aims at guaranteeing the production processes and not the products themselves, by verifying and certifying that each operator in the supply chain (farmers, processors, traders, importers) complies with the correct application of the production rules. After the first EU organic regulation entered in force, several amendments followed, in order to take into account of the dynamic evolution of the organic sector and the new technical-scientific knowledge. Then, after a long-lasting process, in June 2018, a new organic regulation (EU) 2018/848 was published, with the provision of its entry into force from 1 January 2021.

Chapter 3 deals with economics and consumer aspects of the organic aquaculture. The general picture emerging from the analysis of the production process shows higher costs and a differentiated situation in terms of productivity in organic aquaculture, which implies that overall profitability is strongly conditioned by premium prices. However, while a general consensus can be considered for what concerns higher production costs for organic aquaculture, a more differentiated picture emerges for what concerns yields. As a consequence of a lack of knowledge, many consumers do not clearly distinguish between sustainable and organic aquaculture. The two terms are frequently mixed or used synonymously. Nevertheless, consumer perceptions of the product attributes typical of organic farmed fish are mostly in line with current organic aquaculture practices. Organic fish farming is perceived as a natural production method that combines eco-friendliness with fish welfare. Preferences and willingness to pay become of crucial relevance when asking for market opportunities of organic food and seafood.

Chapter 4 discusses aquaculture genetic selection and classic selective breeding programmes. Feed costs normally constitute the highest running expenses in aquaculture; therefore, selection for improved feed efficiency would greatly benefit the industry. Moreover, disease outbreaks pose serious challenges in aquaculture where in many occasions, efficient therapeutic agents are lacking. Selective breeding to prevent the detrimental effects of disease outbreaks is particularly relevant in the case of organic farming where there is restriction in the use of antibiotics and the available therapeutic agents are even more limited. One more element to be analysed is the contribution of potential GxE interactions among different environments and their implications in terms of lower-than-expected genetic gains of the selective breeding programmes carried out.

Chapter 5 emphasizes on hatchery operations in order to overcome the current problems of producing seed (larvae), especially of marine fish, according to organic criteria. Most of the problems are related to the booster used to enrich phytoplankton and zooplankton for larval feeding. Restrictions also exist on the methods to induce spawning, such as the use of hormones and/or the manipulation or ablation of eyestalks in crustaceans. Further research in the enrichment of live prey using only microalgae or combined with other organic products is also essential for a sustainable organic hatchery production.

Chapter 6 considers the aquaculture production systems, its recycling abilities and how technology solutions and science-based production systems and protocols can achieve sustainable and organic aquaculture systems that comply with circular economy principles and maintain healthy aquatic ecosystems. A wide overview of pros and cons in relation to the organic principles of flow-through, RAS, cage and pond systems and IMTA is presented.

Chapter 7 deals with welfare issues and veterinary treatment in organic aquaculture. Farmed fish are exposed to a range of industry practices that may act as chronic stressors, which potentially compromise welfare. The effects of a wide range of aquaculture practices on the stress physiology of fish, including frequent handling, transport, periods of food deprivation, deteriorating water quality, suboptimal stocking densities, fin-clips and environmental enrichment are analysed in this chapter. Effective biosecurity policy that minimizes the risk of introducing and spreading infectious diseases from/to the animals at a facility level, but also of spreading diseases to other sites, is a key factor to reduce the stress of the fish, to prevent disease and veterinary treatments.

Chapter 8 covers aspects of current use of formulated feeds, feed composition, aquafeed technology, sustainable alternatives to common feed ingredients, nutritional physiology and general nutritional principles and product quality in the context of the organic aquaculture. It reviews new knowledge and presents research results to update and modify the criteria and standards for organic aquaculture in relation to nutrition and thus to provide high-quality products for the consumers. Recognition and effective communication on how the organic aquaculture products are developed will work towards greater consumer support.

Research efforts across disciplines worldwide by teams of researches work effectively to meet the future needs of organic aquaculture. As organic aquaculture moves forward, it is important that clear mechanisms are in place to ensure that progress against the outcomes is achieved and monitored.

Bari, Italy Giuseppe Lembo
Volos, Greece Elena Mente

Acknowledgements

We are grateful to colleagues of the EU project OrAqua with whom we have been engaged in several stimulating discussions on a range of issues related to organic aquaculture.

We also gratefully acknowledge the following colleagues who found time to provide constructive comments on the individual chapters: Udo Censkowsky (Bluesensus, Germany), Eduardo Cuoco (IFOAM EU, Belgium), Ian Forster (DFO-MPO, Canada), Cristina Mauracher (Ca' Foscari University of Venice, Italy), Michail A. Pavlidis (University of Crete, Greece), Costas Tsigenopoulos (HCMR-IMBBC, Greece), Jouni Vielma (Luke, Finland) and Manuel Yúfera (ICMAN-CSIC, Spain).

We do express our appreciation to Springer publishers and especially to Alexandrine Cheronet for her valuable editorial services and detailed responsibilities for this book publication.

We welcome reader comments concerning the book.

Contents

1 **Organic Aquaculture: Principles, Standards and Certification** 1
David Gould, Antonio Compagnoni, and Giuseppe Lembo

2 **EU Regulation on Organic Aquaculture** 23
Emanuele Busacca and Giuseppe Lembo

3 **Organic Aquaculture: Economic, Market and Consumer Aspects** 41
Danilo Gambelli, Simona Naspetti, Katrin Zander, and Raffaele Zanoli

4 **Genomics Era on Breeding Aquaculture Stocks** 65
Petros V. Martsikalis, Georgios A. Gkafas, Christos Palaiokostas, and Athanasios Exadactylos

5 **Early Life Stages and Weaning** 79
Alicia Estévez, Nikos Papandroulakis, Mathieu Wille, and Patrick Sorgeloos

6 **Aquaculture Production Systems and Environmental Interactions** 103
Dror Angel, Alfred Jokumsen, and Giuseppe Lembo

7 **Welfare Issues and Veterinary Treatments** 119
Giuseppe Lembo, Pierluigi Carbonara, Andrea Fabris, Amedeo Manfrin, and Walter Zupa

8 **Nutrition in Relation to Organic Aquaculture: Sources and Strategies** 141
Elena Mente, Alfred Jokumsen, Chris G. Carter, Efi Antonopoulou, and Albert G. J. Tacon

Index 189

About the Editors and Contributors

Editors

Giuseppe Lembo is Chairman of COISPA. He has more than 25 years' experience in the field of fishery and aquaculture during which he participated and/or coordinated several EU and national research projects in the field of stock assessment, fishery management, fish telemetry and spatial distribution/dynamics, conventional and organic aquaculture and fish behaviour and welfare.

During his career, he was Member of a number of scientific associations, committees and working groups of several organizations (e.g. Italian Ministry of Agriculture Policy, Italian Society of Marine Biology (SIBM), European Commission, European Aquaculture Society (EAS), Food and Agriculture Organization (FAO), The International Federation of Organic Agriculture Movements (IFOAM)).

He is currently the Chair of the Expert Group for Technical Advice on the Organic Production (EGTOP) of the European Commission.

He was Member of the EU COST Action No. 867 "Welfare on fish in European Aquaculture" (2006–2011). He was Coordinator/Convenor of the IFOAM World Conference on the Organic Aquaculture in Cattolica, Italy (2008), and Uljin, Korea (2011). He took part in the CODEX Committee on Food Labelling for the revision of the Guidelines for the Production, Processing, Labelling and Marketing of Organic Aquaculture (42° session, Rome, 2014, and 43° session,

Ottawa, 2016). He is Editor/Co-editor of the volumes (i) *Studies and Reviews*, selected papers presented at the workshop on biological reference points (n.83, FAO, 2006) and (ii) *Aquatic Telemetry: Advances and Applications*, Proceedings of the Fifth Conference on Fish Telemetry held in Europe (Ustica, Italy, 9–13 June 2003, Rome, FAO/COISPA, 2005); his publications include more than 50 research papers, in peer-reviewed journals, as author/co-author.

Elena Mente is a Professor of aquatic animal nutrition and physiology at the University of Thessaly in Greece and an Honorary Research Fellow at the University of Aberdeen in the UK. She was awarded a Training and Mobility Researchers Postgraduate Fellowship aiming at Ph.D. research in crustacean physiology at the University of Aberdeen, UK. Her research interests lie mainly in aquaculture and nutrition of aquatic animals. Her publications include edited volumes, books and chapters in books (15) and more than 60 research papers in many peer-reviewed journals as co-author/author. She has cooperated with more than 20 research teams in different countries through national, bilateral and EU-funded projects. She currently serves in the editorial team of five well-respected peer-reviewed scientific journals, and she is Associate Editor in the *Journal of Crustacean Biology*. She is actively involved in several research projects supported by the European Commission and by the aquaculture industry. She has also received a number of grants on issues of organic aquaculture and nutrition and reproduction of Norway lobster. She has co-organized the Aquaculture Europe 2011 annual meeting of the European Aquaculture Society in Greece and many workshops related to aquaculture. She was also the Chairman of the Crustacean Society Summer Meeting and the Tenth Colloquium Crustacea Decapoda Mediterranea. She is a Member of the Hellenic Aquaculture Technology and Innovation Platform (HATiP), the sub-group for technical advice on organic production of the European Commission (EGTOP), and served as an Expert (reserve list) at the European Scientific, Technical and Economic Committee for Fisheries (STECF). She is Elected European Governor of the Crustacean Society for 2011–2013. In 2018, she was appointed as a National

Expert to serve in the Pool of Experts by the United Nations to contribute on the regular process for global reporting and assessment of the state of the marine environment for aquaculture issues. She is elected as a Member of the Board of the European Aquaculture Society (EAS) (2012–2018).

Contributors

Dror Angel is an Applied Marine Ecologist at the Recanati Institute for Maritime Studies, University of Haifa. In addition to current research on jellyfish blooms, he has worked extensively on understanding the environmental interactions of aquaculture and is actively searching for sustainable aquaculture models, such as the promotion of omnivore species (e.g. grey mullets) and integrated multi-trophic aquaculture (IMTA). Moreover, he is active in a number of citizen science projects to help bridge the gap between scientists and the public.

Department of Marine Biology and Maritime Civilizations, Charney School of Marine Science, University of Haifa, Haifa, Israel

Efthimia (Efi) Antonopoulou is Biologist with a Ph.D. in Zoology (University of Stockholm, Sweden). She works as Associate Professor (expertise in animal physiology with emphasis on fish physiology) at the Department of Zoology, School of Biology, Faculty of Sciences, Aristotle University of Thessaloniki, Greece. Her research interest is focused on reproductive physiology, nutrition, molecular biology and genomics of fish.

Laboratory of Animal Physiology, Department of Zoology, School of Biology, Faculty of Sciences, Aristotle University of Thessaloniki, Thessaloniki, Greece

Emanuele Busacca is the Regulation Manager at IFOAM EU, the EU branch office of the International Federation of Organic Agriculture Movements (IFOAM Organics International). He is in charge of following EU legislative files, specifically the current and future EU Organic Regulations, and other EU legislations relevant to EU organic operators. Emanuele worked in the past in a Control and Certification Body for Organics being in charge of international organic trade. His educational background includes an M.Sc. in Agriculture and a postgraduate specialization in Quality Management Systems in the Food Industry.

IFOAM-EU, Brussels, Belgium

Pierluigi Carbonara Ph.D., is a Senior Researcher at COISPA and a Fisheries Scientist with 22 years of experience in fishery biology and aquaculture. He is responsible for the Experimental Aquaculture Laboratory (EAL) and the Energetic Metabolism and Telemetry Laboratory (EMTEL). He is the author/co-author of more than 50 scientific publications.

COISPA Tecnologia & Ricerca, Stazione Sperimentale per lo Studio, delle Risorse del Mare, Bari, Italy

Chris G. Carter Institute for Marine and Antarctic Studies, University of Tasmania, Hobart, TAS, Australia. He is Professor of aquaculture nutrition and Head of the Fisheries and Aquaculture Centre, IMAS, at the University of Tasmania. His research is focused on the nutritional physiology of aquatic ectotherms with a focus on developing ingredients and feeds, climate change impacts and sustainable aquaculture systems. Research species include salmonids, snooks, squid and lobsters.

Institute for Marine and Antarctic Studies, University of Tasmania, Hobart, Australia

Antonio Compagnoni since the early 1980s is an organic agriculture passionate advocate and expert. He has been working for relevant international organic certification organizations for over 25 years. Currently he is responsible for international relations of HUMUS network for Italian Organic Agriculture, and he is representing the organic Italian movement at the IFOAM European Union Council.

IFOAM-EU, Rome, Italy

Alicia Estévez Director of Aquaculture Program, Institut de Recerca i Tecnología Agroalimentaries (IRTA), Sant Carles de la Ràpita, Tarragona, Spain. She has a Bachelor of Science in Biology (Complutense University of Madrid, Spain) and Ph.D. in Fisheries (Kagoshima University, Japan) and working in larval rearing and larval nutrition since 1987 and specialized in fatty acid requirements of marine organisms, pigmentation problems in flatfish and larval rearing of new emergent species.

IRTA Sant Carles de la Ràpita, Sant Carles de la Ràpita, Spain

Athanasios Exadactylos is a Molecular Evolutionary Geneticist and holds an academic staff position (Associate Professor) at the Department of Ichthyology and Aquatic Environment, University of Thessaly, focusing his research on aquatic animal and plant genomic footprint.

Department of Ichthyology and Aquatic Environment, School of Agricultural Sciences, University of Thessaly, Volos, Greece

Andrea Fabris (DVM) is the Director and Fish Veterinarian of the Italian Fish Farmers Association (API) and maintains his role as a Veterinary Consultant also carrying out the management of the association with a particular focus on aquaculture sustainability (environmental, social and economic). He is Member of several working groups at the General Direction for Fisheries and Aquaculture of Italian Ministry of Agriculture and Italian Ministry of Health regarding aquaculture, and he is also Member of Aquaculture Advisory Council (AAC).

Director and Fish Veterinarian of Italian Fish Farmers Association – API, Verona, Italy

Danilo Gambelli Ph.D., is Associate Professor at the Università Politecnica delle Marche, Italy, where he lectures in agricultural and environmental economics. He has more than 20 years of experience in researching organic farming, rural development and farm economics.

Department of Agricultural, Food & Environmental Sciences, Università Politecnica delle Marche, Ancona, Italy

Georgios A. Gkafas is a Molecular Evolutionary Biologist holding an academic position (Assistant Professor) at the Department of Ichthyology and Aquatic Environment, University of Thessaly, Greece, focusing on understanding the evolution of biodiversity and the historical context of biogeography for a wide range of aquatic species with a special interest in marine mammals. Research species include cetaceans, marine angiosperms, benthic organisms and osteichthyes.

Department of Ichthyology and Aquatic Environment, School of Agricultural Sciences, University of Thessaly, Volos, Greece

David Gould is Senior Facilitator for IFOAM Organics International and a Global Expert on organic and sustainability frameworks. In addition to working for numerous social and environmental regulatory and labelling schemes, he has been a producer, processor, trainer and consultant to public and private entities for the past 25 years. He chairs the IFOAM Organics International Expert Working Group on Aquaculture.

IFOAM – Organics International, Bonn, Germany

Alfred Jokumsen is Research Coordinator and Senior Advisory Scientist at Aquaculture Section, DTU Aqua, Technical University of Denmark, DTU. He has extended experience in aquaculture: fish farming practices including RAS and environmental efficient production, fish feed and nutrition, organic farming, ECO certification and fish welfare.

National Institute of Aquatic Resources, (DTU Aqua), Technical University of Denmark, Hirtshals, Denmark

Giuseppe Lembo is Chairman of COISPA. He has more than 25 years' experience in the field of fishery and aquaculture during which he participated and/or coordinated several EU and national research projects on topics of stock assessment, fishery management, fish telemetry and spatial distribution/dynamics, conventional and organic aquaculture and fish behaviour and welfare.

COISPA Tecnologia & Ricerca, Bari, Italy

Amedeo Manfrin National Reference Centre for Fish, Mollusc and Crustacean Diseases, Istituto Zooprofilattico Sperimentale delle Venezie, Italy. He is an expert in the sector of fish pathology and welfare with a focus on salmonids and Mediterranean species. He is a partner of several national and international projects. He is Adjunct Professor in infectious diseases, therapy and prophylaxis of fish at Padova University. He is also a Member of international and national working groups on aquatic animal health and welfare.

Fish Pathology Unit, NRL for Fish, Mollusc and Crustacean Diseases, Istituto Zooprofilattico Sper. le delle Venezie, Adria Dria (RO), Italy

Petros V. Martsikalis has an M.Sc. in Food Technology Quality Assurance (University of Reading, UK) and completed his Ph.D. on Population Genetics and Biochemical Analyses in the University of Thessaly, Greece. His research interests include farmed fish stock identification and selective breeding in aquaculture. Research species include rainbow trout, sea bream, sea bass and sharpsnout sea bream.

Department of Ichthyology and Aquatic Environment, School of Agricultural Sciences, University of Thessaly, Volos, Greece

Elena Mente is a Professor of aquatic animal nutrition and physiology at the University of Thessaly in Greece and an Honorary Research Fellow at the University of Aberdeen in the UK. Her research interests lie mainly in aquaculture and nutrition of aquatic animals.

School of Agricultural Sciences, Department of Ichthyology and Aquatic Environment, University of Thessaly, Volos, Greece

Simona Naspetti Ph.D., is Associate Professor at Università Politecnica delle Marche, Italy. She lectures in food economics and marketing. She has participated in several EU-funded research on consumer preferences and has published some seminal papers on organic food consumption.

Dipartimento Scienze e Ingegneria della Materia, dell'Ambiente ed Urbanistica, Università Politecnica delle Marche, Ancona, Italy

Christos Palaiokostas is an Associate Senior Lecturer in the Department of Animal Breeding and Genetics, Swedish University of Agricultural Sciences. His current research involves the incorporation of high-throughput sequencing technologies in the ongoing national Swedish breeding programmes of Arctic charr and rainbow trout (www.vbcn.se), aiming at increasing selection accuracies of complex traits.

Department of Animal Breeding and Genetics, Swedish University of Agricultural Sciences, Uppsala, Sweden

Nikos Papandroulakis is currently Director of Research at IMBBC, HCMR, with more than 25 years' experience in marine finfish aquaculture. He has coordinated or participated in several European and national research projects (32 in total) and is the scientific responsible of the hatchery and the cage farm of the institute.

Hellenic Center for Marine Research, Institute of Marine Biology, Biotechnology and Aquaculture, Heraklion, Crete, Greece

Patrick Sorgeloos (Emeritus Professor of aquaculture at Ghent University, Belgium) has been involved in fish and shellfish larviculture ever since the establishment of the Artemia Reference Center in 1978 at Ghent University in Belgium. Patrick is a strong promoter of international networking in aquaculture and was/is involved with the World Aquaculture Society, the European Commission and the European Aquaculture Technology and Innovation Platform. He was Co-founder of the spin-off company Artemia Systems that is now operating under the name of INVE Aquaculture.

Faculty of Bioscience Engineering Laboratory of Aquaculture & Artemia Campus Coupure – Blok F Reference Center, University of Ghent, Ghent, Belgium

Albert G. J. Tacon is Technical Director of Aquatic Farms Ltd. in Hawaii and works as an International Consultant in Aquatic Feeds and Nutrition. Educated in the UK, Dr. Tacon serves as Editor in Chief of *Reviews in Aquaculture* and also serves on the Editorial Board of Aquaculture Nutrition and Aquaculture Research, respectively. He has over 230 publications dealing mainly on aspects related to aquatic feeds and nutrition and aquaculture development.

Aquatic Farms Ltd, Kaneohe, HI, USA

Mathieu Wille is a Senior Researcher at the Laboratory of Aquaculture & Artemia Reference Center of Ghent University in Belgium. He has over 20 years' experience in different aspects of aquaculture, including live food techniques, broodstock management and larval rearing, specializing in crustacean species (penaeid shrimp, freshwater prawn *Macrobrachium* spp. and mud crab *Scylla* spp.).

Faculty of Bioscience Engineering Laboratory of Aquaculture & Artemia Campus Coupure – Blok F Reference Center, University of Ghent, Ghent, Belgium

Katrin Zander is Senior Researcher at the Thünen Institute of Market Analysis. As an Agricultural Economist by training, her field of expertise now is in consumer research in the wide field of food with a focus on organic and other sustainable production methods. Additionally, she works on the relationship between society and farming by analysing societal expectations and, in doing so, adding to the discourse about future agricultural and food systems.

Thünen-Institut für Marktanalyse, Thünen Institute of Market Analysis, Braunschweig, Germany

Raffaele Zanoli Ph.D., is Full Professor at the Università Politecnica delle Marche, Italy, where he lectures in food marketing and management. He has published more than 100 papers on organic food and farming. He is currently Member of the board of ISOFAR (International Society of Organic Agriculture Research) and Editor of *Organic Agriculture* (Springer).

Department of Agricultural, Food & Environmental Sciences, Università Politecnica delle Marche, Ancona, Italy

Walter Zupa Ph.D., is a Researcher at COISPA. He is involved in the activities of the Experimental Aquaculture Laboratory (EAL) and the Energetic Metabolism and Telemetry Laboratory (EMTEL). He carried out several experiments testing the effects of different housing conditions and diets on the growth performances, energetics, metabolic profile and welfare of several species (e.g. sea bass, sea bream, rainbow trout) in open-flow systems and RAS. He is expert of fish tagging, sensors implantation and wireless communication systems.

COISPA Tecnologia & Ricerca, Stazione Sperimentale per lo Studio, delle Risorse del Mare, Bari, Italy

List of Figures

Fig. 2.1 Flow chart of the EU organic control system 32

Fig. 3.1 Per capita supply of fish (kg/year) for 2013. (Source: Calculated from data from FAOSTAT (2017)) 42

Fig. 3.2 Volumes and value of fish production from aquaculture worldwide. (Source: Calculated from data from FAO – Fisheries and Aquaculture Department – Global Production Statistics 2018) 43

Fig. 3.3 Average world prices of fish from aquaculture and world inflation. (Source: Calculations from data from FAO – Fisheries and Aquaculture Department – Global Production Statistics, and World Bank) 43

Fig. 3.4 Organic aquaculture production, as the main producers* and the annual variation. (Source: Calculated from FiBL data; *95% of total organic world production; Vietnam: 2015 not available) 45

Fig. 3.5 Estimated organic aquaculture production by species in 2016. (Source: Calculated from Lernaud and Willer (2018) and Xie et al. (2013)) 45

Fig. 3.6 Main organic aquaculture producers in Europe (icon sizes are approximatively proportional to volumes of production). (Source: Adapted from Prins et al. (2015) and EUMOFA (2017)) 47

Fig. 3.7 Cost-price structure for selected organically farmed species in selected European countries (average values, % of total costs) (Source: Calculated from Prins et al. (2015)) 48

Fig. 3.8 Make-up of price for farmed fish along the supply chain (% of total unitary price at consumer level): salmon. (Source: Calculated from (Prins et al. 2015)) 53

Fig. 3.9 Make-up of price for farmed fish along the supply chain (% of total unitary price at consumer level): trout. (Source: Calculated from (Prins et al. 2015)) ... 53
Fig. 3.10 Make-up of price for farmed fish along the supply chain (% of total unitary price at consumer level): sea bass/bream. (Source: Calculated from (Prins et al. 2015)) ... 53
Fig. 3.11 Make-up of price for farmed fish along the supply chain (% of total unitary price at consumer level): carp. (Source: Calculated from (Prins et al. 2015)) ... 54
Fig. 3.12 Average additional willingness to pay (in%) for organic fish production compared to the other sustainability attributes indicated. (Source: Adapted from Zander and Feucht (2017)) ... 58
Fig. 3.13 Distribution of consumer willingness to pay (WTP) for 'organically produced' seafood according to country. (Source: own calculations) ... 59

Fig. 5.1 Larval development of gilthead sea bream *Sparus aurata* from hatching (**a**) to 50 days post-hatch (**g**) showing mouth opening (**b**), swim bladder inflation (**c**, 10 dph), flexion (**d**, 20 dph), and tail (**e**) and fins (**f**, 30 dph and **g**, 50 dph) formation. ... 80
Fig. 5.2 Rotifer Brachionus sp. (www.gettyimages.com) ... 82
Fig. 5.3 Development of brine shrimp *Artemia* sp. from cysts to newly hatched nauplii, metanauplii, and adult stage. (www.zootecniadomestica.com) ... 82
Fig. 5.4 Image of a Calanoid (planktonic) copepod *Acartia* sp. (www.CFeed.no) ... 83
Fig. 5.5 Factors affecting feed utilization by marine fish larvae. (Kolkovski 2008) ... 84
Fig. 5.6 Newly hatched larva of salmon showing an opened mouth and pigmented eyes. (www.esacademic.com) ... 86
Fig. 5.7 Larval stages of penaeid shrimp: nauplius, zoea, mysis, and postlarva (with *Artemia* nauplii). (Courtesy: Roeland Wouters) ... 88

Fig. 6.1 Sketch of a recirculating aquaculture system (RAS). The numbers in the figure are referred to (in brackets) in the text. (Source: Billund Aquakulturservice ApS, Denmark) ... 107
Fig. 6.2 Integrated multi-trophic aquaculture (IMTA) flow chart. (Source: IDREEM EU project – final report) ... 114

List of Tables

Table 2.1 Import regimes for organic products 36

Table 3.1 World fish production in 2013 by area [live weight, millions of tons (share, %)] 42
Table 3.2 Organic aquaculture: volumes and share of the main European Union producers (2015) 46
Table 3.3 Main productivity and cost categories for selected species in Europe[a]: relative changes (%) of organic vs conventional aquaculture where not differently specified 49
Table 3.4 Reported relative increases in prices obtained and profitability from organic aquaculture as compared to conventional aquaculture 51

Table 5.1 The effect of *Artemia* replacement 89
Table 5.2 Main differences between hatchery techniques 92

Table 6.1 Advantages and disadvantages of a flow-through organic system and a recirculated aquaculture system 108

Table 7.1 Welfare indicators of fish welfare 121
Table 7.2 Standard metabolic rate (SMR), active metabolic rate (AMR) and scope for activity (SFA) values for sea bass, sea bream, rainbow trout and Atlantic salmon 124

Table 8.1 Acceptable and nonacceptable materials in feeds for organic aquaculture 145
Table 8.2 A selection of similarities and differences to specific rules that apply to the feeding practices of aquatic animals reared in organic aquaculture as they have been defined in various international standards. For each standard/regulation, it is also reported the issue date and the country of origin 151
Table 8.3 Summary of the main current concerns and challenges arising from organic aquafeeds and feeding 174

List of Tables

Chapter 1
Organic Aquaculture: Principles, Standards and Certification

David Gould, Antonio Compagnoni, and Giuseppe Lembo

Introduction

IFOAM Organics International, the global umbrella organization for the organic sector, defines organic agriculture as "a production system that sustains the health of soils, ecosystems and people". It relies on ecological processes, biodiversity and cycles adapted to local conditions, rather than the use of inputs with adverse effects. Organic agriculture combines tradition, innovation and science to benefit the shared environment and promote fair relationships and a good quality of life for all involved (IFOAM 2014).

Historically, systems and products using the term "organic" (or its equivalent in other languages, e.g. "biologique", "ecológico", etc.) have indeed focused on the health and stewardship of soils as their foundation for production. Soil gives rise to the crops and livestock grown as primary products, which then may be processed further before they reach the final consumer. These are in essence terrestrial, soil-based systems; without the basis of soil and without good soil stewardship, organic production of crops and livestock in a terrestrial setting does not exist.

Organic systems are rooted in four Principles of Organic Agriculture: health, ecology, fairness and care. "The Principles apply to agriculture in the broadest sense, including the way people tend soils, water, plants and animals in order to

D. Gould (✉)
IFOAM – Organics International, Bonn, Germany
e-mail: d.gould@ifoam.bio

A. Compagnoni
IFOAM-EU, Rome, Italy
e-mail: acompagnoni@database.it

G. Lembo
Stazione Sperimentale per lo Studio delle Risorse del Mare,
COISPA Tecnologia & Ricerca, Bari, Italy
e-mail: lembo@coispa.it

G. Lembo, E. Mente (eds.), *Organic Aquaculture*,
https://doi.org/10.1007/978-3-030-05603-2_1

produce, prepare and distribute food and other goods. They concern the way people interact with living landscapes, relate to one another and shape the legacy of future generations (IFOAM 2014)."

In contrast to agriculture, aquaculture does not, as a rule, have soil stewardship at its heart, although production of certain aquatic species involves caring for the soil that either abuts or is submerged under a terrestrially located aquatic system such as a pond, river or estuary. Nonetheless, the organic principles may be applied to aquaculture systems, and there are good reasons to do so.

Globally, aquaculture continues to play an increasingly important role in the provision of animal protein in human diets. Yet, aquaculture production systems can put increasing strain on wild aquatic systems by depleting feed sources for wild populations or polluting surrounding areas with contaminated effluent.

The goal of the organic sector on the other hand is to positively impact aquaculture production through the establishment of ecologically integrated systems that preserve the natural environment, maintain or enhance biodiversity, respect animal welfare and yield high-quality, healthy products. For example, aquaculture feeds produced through organic methods avoid the use of toxic substances that can have negative environmental and human health effects. Furthermore, organic health treatments do not use antibiotics and other drugs that can carry over into the human diet or encourage pathogenic resistance to such cures. While organic standards and certification currently cater to a growing market of goods specially labelled as organic, the sector also aims at contributing to the improvement of production practices more generally. Organic practices offer solutions and benefits for all kinds of agriculture and aquaculture operations regardless of the market claims involved (Arbenz et al. 2016).

The organic label distinction nonetheless remains an important vehicle for facilitating consumer choice, and consumer demand for organic goods increases annually (Willer and Lernoud 2017). Consumers who buy organic products do so for common reasons, including the quality of the final product both in terms of taste and expectation of lower or non-presence of toxic residues, concern for the environment and animal welfare. The year-on-year increase in demand is made possible in part because consumers have confidence in the credibility of the claims made on organic products. The guarantee has two main complementary facets: the rigour of the practices mandated by the standard (private sector standard or government regulation) and the rigour with which those practices are verified as having been carried out. Consumers must believe that the standard's requirements are meaningful and that the checking is done in a serious and trustworthy manner.

It is thus important that when extending the scope of organic products beyond the more "traditional" range of agricultural goods, namely, crops, livestock and products derived therefrom, into the realm of aquatic systems, consumers retain their sense that these aquaculture products are compatible with what they expect "organic" to mean. In fact, some people, a minority by all accounts, deny that organic aquaculture is even a valid concept, usually citing either the lack of soil as a central focus. The fundamental question of whether or not the organic sector should include aquaculture in its scope of activities was debated through public consultation and then put to a vote by the General Assembly of IFOAM Organics

International in 2017, with the result in favour of inclusion. The "translation" of the Principles of Organic Agriculture to aquaculture systems thus requires some adjustment of emphasis or re-contextualization in order to assure that organic aquaculture requirements reflect the intention of the Principles to the greatest degree possible.

Aquaculture has been practised in certain cultures and locations for many generations, centuries if not longer, integrated with certain agricultures, e.g. in parts of Asia and in natural estuaries in various places. The increase of aquaculture however to volumes of product that reflect its significant impact on total animal protein consumption by humans is due to more intensive systems and technological innovations in recent decades than has been done before. These newer, higher-producing systems are often done with practices that are not as clearly or wholly compatible with the Principles. Stakeholders have had to weigh positive and negative considerations and make compromises in the spirit of meeting the needs, desires and expectations of producers and consumers alike.

Ongoing analysis of the effects of organic aquaculture systems will help reveal how well such production systems meet the expectations of both consumers and producers in terms of environmental performance and productivity, product quality and nutritional value and economic viability of organic aquaculture enterprises.

Organic Principles Applied to Aquaculture

The Principles form an integrated, holistic conception. The main aspects of aquaculture practice, e.g. site location and design, breeding, feeding, species health and welfare and product handling and quality, do not manifest singly, but rather in conjunction with each other and therefore impinge on multiple principles at the same time, as can be illustrated through the following discussion of each Principle:

Principle of Health

> *Organic agriculture should sustain and enhance the health of soil, plant, animal, human and planet as one and indivisible.*
>
> *This principle points out that the health of individuals and communities cannot be separated from the health of ecosystems – healthy soils produce healthy crops that foster the health of animals and people.*
>
> *Health is the wholeness and integrity of living systems. It is not simply the absence of illness, but the maintenance of physical, mental, social and ecological well-being. Immunity, resilience and regeneration are key characteristics of health.*
>
> *The role of organic agriculture, whether in farming, processing, distribution or consumption, is to sustain and enhance the health of ecosystems and organisms from the smallest in the soil to human beings. In particular, organic agriculture is intended to produce high-quality, nutritious food that contributes to preventive healthcare and well-being. In view of this, it should avoid the use of fertilizers, pesticides, animal drugs and food additives that may have adverse health effects* (https://www.ifoam.bio/en/organic-landmarks/principles-organic-agriculture).

In terms of aquaculture, the main implications of this Principle devolve to (i) the health of the species in question through proper nutrition and (ii) the nutritional quality of the product sold for human consumption. Because in aquaculture systems carnivorous fish often receive a portion of their diet in plant-based form, there is a reason to be concerned that the plant-based portion may provide a different balance of nutrients, which may in turn affect the health of the species and the nutritional profile of the final product. For example, salmonids require essential fatty acids in their diet. Being deprived of a more "natural" balance as they might get in open waters could lead to poor animal health, a practice which would be against organic principles and standards. Furthermore, fish with lower-than-normal omega-3 fatty acid levels might fall short of consumer expectations, a danger toward the credibility and market success of an organic label.

While the stewardship of the environment in which aquaculture species are raised is important to their health, the soil loses its primacy in aquaculture in contrast to agriculture operations. Nonetheless, the aquatic environment has significant implications for the safety of the food products involved. For example, naturally occurring bacterial blooms can make seafood products unsafe to eat, and waters contaminated otherwise by biological or chemical agents can pose risks that organic standards purport to minimize. Furthermore, postharvest handling is especially critical for aquatic animal products to assure that spoilage is avoided and the freshness and organoleptic qualities of these highly perishable products are maintained.

Principle of Ecology

> *Organic agriculture should be based on living ecological systems and cycles, work with them, emulate them and help sustain them.*
>
> *This principle roots organic agriculture within living ecological systems. It states that production is to be based on ecological processes and recycling. Nourishment and well-being are achieved through the ecology of the specific production environment. For example, in the case of crops, this is the living soil; for animals it is the farm ecosystem; and for fish and marine organisms, it is the aquatic environment.*
>
> *Organic farming, pastoral and wild harvest systems should fit the cycles and ecological balances in nature. These cycles are universal but their operation is site-specific. Organic management must be adapted to local conditions, ecology, culture and scale. Inputs should be reduced by reuse, recycling and efficient management of materials and energy in order to maintain and improve environmental quality and conserve resources.*
>
> *Organic agriculture should attain ecological balance through the design of farming systems, establishment of habitats and maintenance of genetic and agricultural diversity. Those who produce, process, trade or consume organic products should protect and benefit the common environment including landscapes, climate, habitats, biodiversity, air and water* (https://www.ifoam.bio/en/organic-landmarks/principles-organic-agriculture).

The aquatic environment is specifically mentioned in this Principle, giving further credence to the idea that organic aquaculture indeed fits within the concept of what "organic" encompasses. In line with the Principle, the integration of the production system with nature and its environmental impact are of significant concern with regard to practically every main facet of activity.

In terms of breeding and species chosen, organic operators should be wary of the introduction of any exotic or potentially invasive species to new locations. The chance of escape of such organisms from a contained aquaculture operation into open waters is an ongoing concern. Even within the boundaries of an established system, the possibility of a target species under production should not be allowed to overwhelm the biodiversity that otherwise exists.

Feeding species under production, in particular carnivorous species, remains an ongoing challenge for organic aquaculture operations. In the context of the Principle of Ecology, carnivorous species so far have to rely to a significant extent on fish and/or fish trimmings from wild fisheries. Depletion of wild stocks of fish caught in order to feed fish in aquaculture operations can threaten the survival of the species that normally prey on them, disrupting natural marine food chains. Fishmeal derived from trimmings of fish caught for human consumption is a widely used source, but this is not often is not a sufficient supply in quantity and/or quality to maintain the cultured population. Furthermore, generally speaking, organic standards require that livestock raised to be organic products in the marketplace be fed with organic feed, which the wild fishery sources are by definition not. On the other hand, raising fish through organic aquaculture operations and then using them to feed other fish populations, while theoretically possible, proves so far not to be an economically viable option. The organic sector is pressing for the promotion of other protein sources, such as the cultivation of insects or microalgae, in order to use these products as a component of fish feed rations; the degree to which this may alleviate the aforementioned challenges remains to be seen.

One of the most controversial issues in the organic aquaculture debate has been about the recirculation or reuse of water in organic systems. With respect to the Principle of Ecology, there are advantages and disadvantages that stakeholders have had to weigh. The responsible use of water is an increasingly critical issue worldwide for all kinds of production systems. In most conventional operations that reuse water, recirculating aquaculture systems (RAS) are used. In several aspects, these systems pose ecological advantages: RAS have comparatively lower water consumption compared to other systems, and it is easier to disinfect and clear water. RAS increase the ability to recycle water and use effluent and nutrients productively. It is possible to design systems that are in contact with water, but save water and avoid pollution; much of the outcome in this respect depends on good management practices. A big disadvantage is that energy use by RAS is very high, which is a negative attribute especially when derived from non-renewable or greenhouse gas-emitting resources or otherwise having negative impact on the environment or food supply. Furthermore, RAS entail an almost total disconnection with natural environment, which is not in line with the Principle of Ecology (nor with the Principle of Fairness as relates to animal welfare, see below). This has been the overriding idea that has led the organic sector to reject RAS that are not integrated into the natural environment. At its 2017 General Assembly, IFOAM Organics International reached the following resolution regarding system boundaries for organic aquaculture systems:

> *Organic aquaculture may include an environmentally integrated recirculation system only if it is primarily based on and situated in a natural environment. It does not routinely rely on external inputs such as oxygen, allows the raised species to spend the majority of their lives in outdoor facilities and preferably uses renewable energy.* (IFOAM 2017)

With such connection to a natural environment, issues of water quality and effluent remain important concerns in terms of the ecosystem in which any given operation is situated.

Principle of Fairness

> *Organic agriculture should build on relationships that ensure fairness with regard to the common environment and life opportunities.*
>
> *Fairness is characterized by equity, respect, justice and stewardship of the shared world, both among people and in their relations to other living beings.*
>
> *This principle emphasizes that those involved in organic agriculture should conduct human relationships in a manner that ensures fairness at all levels and to all parties – farmers, workers, processors, distributors, traders and consumers. Organic agriculture should provide everyone involved with a good quality of life and contribute to food sovereignty and reduction of poverty. It aims to produce a sufficient supply of good-quality food and other products.*
>
> *This principle insists that animals should be provided with the conditions and opportunities of life that accord with their physiology, natural behaviour and well- being.*
>
> *Natural and environmental resources that are used for production and consumption should be managed in a way that is socially and ecologically just and should be held in trust for future generations. Fairness requires systems of production, distribution and trade that are open and equitable and account for real environmental and social costs* (https://www.ifoam.bio/en/organic-landmarks/principles-organic-agriculture).

The primary considerations with respect to this Principle relate to (i) the health and welfare of animals raised under aquaculture systems, including respecting their natural behaviours and (ii) how fair the production rules are with respect to the economic viability of aquaculture enterprises and the livelihoods of organic producers. Each of these considerations relates to multiple facets of aquaculture production systems, with a mix of pros and cons.

With respect to animal health and welfare, system design and living conditions have great impact. RAS totally disconnected from natural conditions violate the Principle of Fairness by sheer disrespect of the animal's intrinsic right to exist and exhibit its "natural behaviour" under anything close to the conditions under which it evolved. Furthermore, the economic pressures of operating RAS demand higher stocking densities in order to produce greater amounts of product and generate adequate return on investment. Higher stocking densities also carry risk of accelerating spread of disease, should it arise. On the other hand, RAS offer the possibility of controlling water quality against harmful extremes, as well as a degree of certainty to prevent escapes and protect against pathogens. Finally, consumer perception about organic aquaculture products is highly important to the overall well-being of the sector. Many consumers emphasize their priority that organic fish are, and should be, raised in some kind of "natural environment".

From the standpoint of economic considerations for organic producers, RAS can outcompete less intensive aquaculture systems, i.e. those that are more integrated with nature, since it has higher production rates per unit. Other economic pressures include feed requirements of organic standards, which if overly strict or "purist" eliminate the possibility of using certain feed supplements, e.g. certain essential amino acids produced through fermentation, that could allow a higher portion of plant-based feed. In terms of sources of stock, the availability, or lack thereof, organic origin is significantly lacking in certain species; too strict requirements for organic juveniles thus so far simply make organic production unfeasible.

Principle of Care

> *Organic agriculture should be managed in a precautionary and responsible manner to protect the health and well-being of current and future generations and the environment.*
>
> *Organic agriculture is a living and dynamic system that responds to internal and external demands and conditions. Practitioners of organic agriculture can enhance efficiency and increase productivity, but this should not be at the risk of jeopardizing health and well-being. Consequently, new technologies need to be assessed and existing methods reviewed. Given the incomplete understanding of ecosystems and agriculture, care must be taken.*
>
> *This principle states that precaution and responsibility are the key concerns in management, development and technology choices in organic agriculture. Science is necessary to ensure that organic agriculture is healthy, safe and ecologically sound. However, scientific knowledge alone is not sufficient. Practical experience, accumulated wisdom and traditional and indigenous knowledge offer valid solutions, tested by time. Organic agriculture should prevent significant risks by adopting appropriate technologies and rejecting unpredictable ones, such as genetic engineering. Decisions should reflect the values and needs of all who might be affected, through transparent and participatory processes* (https://www.ifoam.bio/en/organic-landmarks/principles-organic-agriculture).

The organic sector considers itself a promoter of innovation and strives to develop techniques and technologies through research and practical experimentation that are appropriate for organic systems. In the aquaculture context, innovations have been considered in several key areas, including feed formulation, as mentioned above with respect to insect production and fermentative sources of amino acids and other supplements, and water management, such as through alternative energy sources to move and reuse water. Because the organic sector has such a strong history and consumer reputation that animal breeding or rearing does not happen via use of hormones, breeding and production of juveniles of certain species remains elusive and bears additional research.

Prevention of adverse impacts of exotic and/or invasive species via their introduction or through escapes also falls within the scope of consideration of the Principle of Care. The overall impact of aquaculture on freshwater and marine life remains an ongoing consideration and merits monitoring on a continual basis.

Organic Aquaculture Standards and Regulations Around the World

The term "standard" implies a set of practices and related performance requirements that must, in order to relate to a credible claim on a product or service, be coupled with some kind of certification or other assurance mechanism. The term "regulation" often includes the practice requirements as well as the control scheme, plus potentially other kinds of surveillance or data collection aspects. For the sake of this discussion about the practice and performance requirements involved in organic aquaculture, we shall use the terms "standard" and "regulation" interchangeably to refer to the practices operators must follow in order to comply with actual production requirements.

Standards serve two main purposes: (i) they describe a set of practices and/or outcomes that lead to a certain quality or classification of products and services, which can be used in and of themselves to guide production activities and (ii) as the basis for a verification of the described practices for the purposes of measuring performance or for the purposes of guaranteeing market claims on products made in accordance with the standard. In terms of "organic aquaculture", the standards do serve both of these purposes. Like in other "organic" claims in the market, the purpose is to distinguish those products from others which have been produced using practices and/or materials that are not allowed by the standards, the implication being that the organic goods reflect certain values, practices and qualities that certain consumers deem desirable.

Governments may set standards as a legal requirement for making organic claims in the marketplace; hence they become regulations. Private sector organizations may also create standards and operate assurance schemes as a service to a given sector based on geographical and/or activity scope. In the organic market sector, both types are essentially labelling schemes whose main purpose is to enable identification of products as organic based on a verification mechanism, usually third-party certification.

Whereas private schemes may operate potentially anywhere, there is a demand from operators or consumers and government permission or licence to do so; government regulations are enforced primarily within their respective legal jurisdictions, their focus being on protecting their citizens as consumers via a credible guarantee about the product's label claims. If ingredients however originate in regions outside the government's jurisdictional boundaries, they may still impose the requirements of their regulation on those relevant operators. This imposition occurs either by requiring direct compliance with the government's own standard or through their recognition of equivalence of a complementary assurance scheme, public or private, that covers those preceding links in the chain.

Some government regulations permit the operation of private schemes within their boundaries and allow those private claims in the market, while others do not. There are various permutations on this theme, which also change over time as government regulatory schemes make revisions. For example:

- The European Union's regulation for organic aquaculture operates as a "base" regulation for organic aquatic products in the community's market but allows other private standards schemes within its borders to also operate their own aquaculture standards and certification programmes with their own labels, as long as these at least comply with the EU regulation.
- The EU regulation has allowed foreign, so-called third country, organic certifications to demonstrate and receive recognition of equivalence so that products from those countries may enter the EU market with organic labelling. Production practices are deemed equivalent, but final labelling for the EU market complies with the EU regulation's labelling requirements. The EU recently approved a revised regulation however, and equivalence by private schemes will no longer be an accepted avenue for importing organic goods. Only country-to-country equivalence arrangements will receive consideration for continued equivalence and those pre-existing may still be renegotiated.
- The US Department of Agriculture's National Organic Program (NOP) does not currently cover aquaculture. Some aquaculture products are labelled as organic in the US market, but without any reference to the USDA NOP. The development of organic aquaculture standards remains on USDA's work plan; should they come into force, under existing policies only regulations included in a US country-to-country agreement will be recognized as equivalent and gain entry to the use market; otherwise direct compliance with the NOP rules will be required.
- The China organic regulation only allows products in the Chinese market if they fully comply with the regulation. Private schemes are not recognized.
- In February 2018 Canada published its organic aquaculture standard, which in and of itself does not specify the allowance of imported organic aquaculture products or those certified by bodies not accredited to the Canada Organic Regime. These aspects presumably will fall under the broader terms of that country's recognition agreements with other organic programmes.

Two main international efforts, namely, Codex Alimentarius and IFOAM Organics International, have strived to create a globally usable organic aquaculture standard aimed at enabling equivalence among standards and concomitant reduction of trade barriers that are caused by relatively minor differences among them. The Codex standard creation process was suspended at the 43rd meeting of the Codex Committee on Food Labelling (May 2016, Ottawa, Canada) after a failure of country delegations to agree on technical requirements for the standard, the issues of contention running the full gamut of difficult topics as outlineds at the beginning of this chapter. In 2017 the IFOAM process broke through a years-long stalemate regarding system boundaries and recirculation/reuse of water, paving the way for including aquaculture as part of the global IFOAM Standard for Production and Processing, which aims to be globally applicable, allowing for regional variations based on the degree of sector development and technical limitations due to such factors as climate and day length.

Achieving actual global equivalence among organic standards, aquaculture and otherwise, remains elusive despite relatively small differences among most respective sets of requirements. Bilateral arrangements are becoming more common, with explorations into multilateral agreements also starting (Willer and Lernoud 2017). The IFOAM Family of Standards (https://www.ifoam.bio/en/ifoam-family-standards-0) serves as a working model for multilateral equivalence among private organic standards and governmental organic regulations and currently includes and recognizes approximately 60 different schemes, but has until now not included aquaculture in its scope. The benchmark for determining equivalence, namely, the Common Objectives and Requirements of Organic Standards (COROS), has not covered aquaculture, a situation that will change once the IFOAM Standard also includes aquaculture in its scope.

Globally organic aquaculture standards, although having existed in certain contexts many years, at least since 2000, are still not very broadly applied compared to agriculture, less than 1% of the organic total area (Willer and Lernoud 2017), and are a relatively new work area for many entities in the sector. As the foregoing discussion described, this slow development is reflective of the technical challenges of organic aquaculture practices, especially when taken through the lens of terrestrial agriculture systems. Expectations for raising terrestrial animals and aquatic ones are not, and need not be, necessarily fully congruent. Resolving the differences between these two lines of thinking, and production, has been a long and slow journey that is gaining momentum, with the recent decisions taken by IFOAM and the further evolution of private and government standards accelerating the process. The challenges in resolving key issues in the standards have also had implications for economic viability of certain kinds of organic aquaculture operations; finding adequate compromises remains a challenge in some cases, and there is correspondingly a call among certain practitioners for increased opportunity to innovate within an organic regulatory context.

Data collection on organic aquaculture is incomplete; certain countries either do not readily provide statistics on organic producers, and non-certified producers are so far generally not counted. The largest producer by volume of certified organic aquaculture products is by far China, which includes aquaculture as part of its national regulation (The National Standard of the People's Republic of China, GB/T 19630.1–2011) and comprises almost 80% of the recorded world total. The next most significant volume of certified production comes under the European Union regulations, covering most of the remaining 20% of the global total. Other government regulations include Brazil, Argentina and Canada, each with still quite low volumes. The East African Organic Products Standard, a regional standard owned by the East African Community, is currently under revision and aims to include a new chapter on aquaculture. As stated above, USDA and its National

Organic Standards Board has had the drafting of an aquaculture section of the NOP regulation on its work plan for some years.

Private organic standards schemes include, among the most active, Naturland (Germany-based but active globally), Soil Association (UK), Organic Agriculture Certification Thailand, and Krav (Sweden). Others include (https://www.ifoam.bio/en/ifoam-accredited-certification-bodies) NASAA (Australia-based), Australian Certified Organic, Instituto Biodinamico (Brazil), Organic Food Development Center (China), Hong Kong Organic Resource Centre Certification Limited, JONA (Japan), AsureQuality Limited (New Zealand-based), CCPB (Italy-based) and Doalnara Certified Organic Korea. Volumes of this latter set are still quite low; this list is however not exhaustive.

Scope and Content of Organic Aquaculture Standards

Organic aquaculture covers a full spectrum of aquatic species, including micro and macro algae, zoo plankton and animals (all kinds). To date, all standards are in unison in declaring that wild fish capture does not qualify for an organic certification or market seal. The East African Organic Products Standard revision currently proposes allowing wild fishery products into the scope of organic. In the United States, controversy over this issue blocked progress of aquaculture standards development even prior to the advent of the National Organic Program, with wild salmon fisheries in the state of Alaska arguing that their wild product merited an organic label as much as did a farmed version (Gould and Kirschenmann 2006).

Standards requirements, in order to be credible, must be verifiable. Since the Organic Principles are a basis of thought about what organic production and products should ideally represent, these Principles should therefore be "translated" into a set of required practices, i.e. an organic standard, that fully covers them. The Principles envision a more ideal world where humanity manifests the full socially, ecologically, economically and culturally accountable spectrum of behaviours. Their integrated, holistic narrative should then be dissected into enough detail to guide producers toward these desired outcomes, which in essence paint a picture of true sustainability for agriculture-based production.

Despite this ideal, all of the organic government regulations, and, generally speaking to a lesser degree, all of the private standards, fall short of fully covering the Principles. The history behind such compromises made in the evolution of organic standards reveals a complex set of considerations and circumstances, having to balance meaningfulness with achievability with the overall intention to create the maximum positive impact. A standard that is too difficult to achieve because its demands are too high does not reach enough operators or consumers to have much impact globally; on the other extreme, too easy requirements may not bring enough collective change even though more operators participate. Organic aquaculture standards are no exception and follow the same basic range of practice considerations as their terrestrial counterparts.

Seeing the gap between organic standards on the whole and the ideal of the Principles, in 2012 IFOAM and its allies formed the Sustainable Organic Agriculture Action Network – SOAAN (IFOAM 2011. General Assembly Motion 57, https://www.ifoam.bio/sites/default/files/minutes_in_action.pdf), a think tank with the intention to refresh the narrative and bring the message of the organic movement, i.e. the Principles, closer to message of the market (the guarantees brought through certification). This initiative reflected the IFOAM membership's awareness that the impacts of the organic sector over the preceding decades had helped give rise to several competing social/environmental standards and labelling schemes and that the level of maturity of the organic market could withstand further demands of performance by its own expectations.

The first phase of the SOAAN think tank's work (2012–2013) produced the Best Practice Guideline for Agriculture and Value Chains (IFOAM, Best Practice Guideline for Agriculture and Value Chains, 2013, http://www.ifoam.org/sites/default/files/best_practice_guideline_v1.0_ratified.pdf), essentially a benchmark document describing in detailed terms the full spectrum of topics to be considered for a complete treatment of sustainability by organic operators along the value chain, from primary production all the way up to the final consumer.

While deliberately not a standard or set of mandatory requirements per se, the Guideline describes in a more practical and detailed manner how to actually manifest the Principles. Even though the fuller scope of practices described by the Guideline exceeds the demands of organic regulations and, to a lesser degree, most private standards, many certified operators' performance already include at least some of the additional practice areas described. Private standards tend as a rule to encompass a broader set of sustainability dimensions, e.g. societal, ecological, economic, cultural and accountability, compared to government regulations. Their users and stakeholders are often more self-selecting, hold more values in common and are smaller in numbers, making for a less controversial and therefore streamlined process in creation and revision.

Certain aspects of production remain challenges for organic producers and consumers to execute under current market and policy conditions. Dependence on non-renewable energy sources, closing nutrient loops in production systems, use of packaging materials and waste reduction, limitations on processing methods that remove nutritive value of the final product and assuring a fair price for farmers and farm labourers are prominent examples. In organic aquaculture chains, the set is similar, with the aspects needing most improvement varying with a somewhat different focus, for example, on water over soil, animal welfare under production systems that may essentially "unnaturally accelerate" the domestication of certain species of fish, compared, for example, to the millennia-old practice of cattle husbandry, and considerations regarding maintenance of quality of relatively more quickly perishable final products.

While making improvements to organic operations carries its own inherent technical challenges, the main inhibiting factor is economic and political: organic products must compete against conventionally produced goods in the market. The additional requirements imposed by organic standards, let alone the full spectrum described by the Best Practice Guideline, pose a greater economic burden on producers and consumers. Simply put, policy and market conditions are such that it is economically more advantageous to exploit the environment in ecologically unsustainable ways and to detract the livelihoods of enterprises and labourers in the value chain than it is to do the opposite. While the organic market is now heavily regulated, and almost 90 countries with official organic rules, the market plays by its own rules, which are not founded on the Principles, - rather it is far from it. Organic products do as a rule sell for a higher price than conventional ones, and the organic market share keeps growing worldwide (Willer and Lernoud 2017). Nonetheless, desires to keep improving organic market requirements toward better fulfilment of the Principles have limits under current conditions, with consumer tolerance for the price increase being only so elastic. The cry from the organic sector is not that organic products cost too much, but rather that conventional ones are too cheap (http://www.hortidaily.com/article/27183/Organic-isnt-too-expensive,-conventional-is-too-cheap).

Seeing this inequity, SOAAN convened a second phase of work in 2014 to better define the socio-political as well as the technical conditions to enable producers and consumers to adopt best practices. The result culminated in 2015 with the official global launch of Organic 3.0, a term initially coined by a group of German stakeholders as far back as 2009 (Arbenz et al. 2015), marking a new phase of the organic movement's evolution. Organic 3.0 calls for continuous improvement by the sector in order to increase performance and global impact by innovating in terms of stakeholder engagement and participation among value chain actors, consumers and policymakers both inside the organic sector and otherwise. This conception is described as having six integrated features that act on all of agriculture-based production and consumption[1]:

One of the features of Organic 3.0 is to bring a fuller cost accounting of production and consumption to the fore in order to reflect true value and fair pricing of goods and thereby place organic systems on an even playing field with conventional ones. Until there is more substantive progress in this regard, however, the organic sector remains constrained by current market competition conditions.

Elements of Organic Aquaculture Standards

Notwithstanding variation among the technical details of different organic aquaculture standards, they all deal more or less with a common set of aspects, including the following categories and their respective key considerations, which have been developed by technical experts and heeding studied consumer expectations.

---

[1] ibid.

System Design and Location

- Avoidance of contamination of the aquatic production operation from outside sources of pollution.
- Adequate isolation from non-organic production systems and segregation of product all along the supply chain.
- Caution against introducing exotic species to the host environment.
- Control of effluent and nutrient deposition into the environment by the organic system so as to minimize negative impacts.
- Degree of integration into the natural environment so as to optimize effects on biodiversity and to provide as natural environment for fostering animal welfare, while controlling escapes.
- Protection of vulnerable wetland ecosystems (e.g. deforestation of mangroves for organic shrimp aquaculture is prohibited).
- Earthen ponds for certain species (liners are not allowed in grow-out ponds).
- Conditions for use and reuse of water and for maintaining water quality, including flow rates, oxygen levels, energy sources for recirculation or temperature regulation. According to IFOAM, organic aquaculture may include an environmentally integrated recirculation system only if it is primarily based on and situated in a natural environment. It may not routinely rely on external inputs, such as oxygen, and must allow the raised species to spend the majority of their lives in outdoor facilities.

Conversion to Organic Aquaculture

- Adequate cleaning of existing systems prior to use for organic production typically ranges from 3 to 12 months.
- Adequate time to establish and verify organic practices is in place, typically a full life cycle of the target species, sometimes a smaller fraction, but generally not less than 2/3 of the species' life.
- Adequate conversion of existing non-organic stocks before they can be used for organic breeding or sold as organic product, typically at least 3 months before allowing for breeding, and at least 2/3 of the life cycle, or a correspondingly high percentage of weight gain, for sold product.

Sources of Stock, Breeds and Breeding

- Preference for locally adapted species
- Breeding using organic stock, seed or juveniles, in preference to non-organic stock, with the entire life cycle of the harvested product having been grown

under organic conditions, although availability of organic juveniles or seed for certain species or in certain regions remains a challenge, so some tolerance for using non-organic sources and heeding a conversion period for juveniles
- Prohibition of polyploidy, artificial hybridization or non-manual means of creating monosex populations
- Prohibition on use of hormones
- Restrictions on artificially lengthening the daylight period

Algae and Microalgae Production

- Restrictions on nutritional inputs to the production system
- Control of harvesting methods to assure long-term system productivity

Production of Molluscs

- Restrictions on nutritional inputs to the production system
- Careful integration of molluscs with the rest of the production environment, i.e. appropriate polycultures and cultivation locations, e.g. bottom cultivation versus other scenarios

Feeding and Nutrition of Aquaculture Animals

- Efficient use of feed, i.e. minimized loss to the environment.
- Restrictions on sources of feed especially for carnivorous species, which must rely on a significant portion of their diets from wild fishery sources, and hierarchies for choosing sources, e.g. (i) fish meal and fish oil from organic aquaculture trimmings; (ii) fish meal and fish oil derived from trimmings of fish, crustaceans or molluscs already caught for human consumption in fisheries that have been certified sustainable under a well-recognized scheme; (iii) fish meal and fish oil and ingredients of fish origin derived from whole fish, crustaceans or molluscs not caught for human consumption in fisheries that have been certified as sustainable under a well-recognized scheme; and (iv) organic feed materials of plant origin. The determination of what qualifies a fishery as sustainable can devolve to national or international governmental standards, or to private schemes, depending on the aquaculture standard in question.
- Restrictions on fish meal and fish oil percentages in the fish diet.
- Restrictions on feed supplements and additives in accordance with lists established as part of the given standard.

- As noted previously, the possibility to use alternative protein sources, such as microalgae or insects produced under organic methods, is a new work area that may bring helpful innovations to the topic of aquaculture feed and nutrition.

Health and Welfare

- Emphasis on cultural practices and prevention, e.g. good system design, choice of species, optimum nutrition, minimized stress from problems related to stocking density appropriate to the species, contaminants or swings in water conditions, e.g. temperature and oxygen levels.
- Defined stocking densities per species.
- Routine cleaning of systems and fallow periods between production cycles as appropriate.
- Restrictions or prohibitions on mutilations.
- Appropriate veterinary treatments as necessary, with limits on frequency in order to label a product as organic. Prophylactic uses of veterinary drugs are prohibited, as are synthetic hormones and growth promoters.
- Use of antibiotics leads to decertification of treated animals.

Harvest, Transportation and Slaughter

Many government regulations do not address these aspects, but most private standards do.

- Minimized stress pre-slaughter, via control of the environment and transport medium
- Minimized suffering during slaughter, usually by rendering animals insensate
- Maintenance of product quality post-slaughter via hygiene and temperature control

Processing and Labelling

- Adequate segregation from non-organic product streams
- Transparent traceability system in place
- Restrictions on additives, processing aids and minor non-organic ingredients, based on lists established as part of the given standard
- Label indications in compliance with the standard and other relevant statutes

Guarantee Systems, Certifications, Market Access and Labelling

Like all other organic standards and regulations, in terms of consumer guarantee and protection, the guarantee, or assurance system, focuses on assuring certain basic facets of the standard have been met for the relevant products in the marketplace, namely:

(i) The practices employed by the producers involved, i.e. producers, processors and other handlers, are known and verified, i.e. the who, what, when, where and why.
(ii) Any input materials used are only those allowed by the standard.
(iii) Operators take all required steps to avoid contamination of the production environment and the products.
(iv) Documentation is maintained to attest to the veracity of operators' stated activities including the traceability of products along the chain of custody.
(v) Market claims and labelling are accurate and follow applicable laws.

Certification

Virtually all third-party organic certifications are based on conformity with ISO Guide 17065: Conformity Assessment – Requirement for Bodies Certifying Products, Processes and Services. In the organic context these, there are process-based certifications, i.e. the final product claim is not a guarantee of actual product content. The implication is that the processes (practices) followed by the operators along the chain lead to products with a degree of certainty that they possess specific attributes, e.g. have been produced without deliberate use of genetic engineering, harmful pesticides and hormones and have respected aspects of animal welfare, biodiversity and, in the case of certain private standards in particular, social justice issues.

The certification process is more or less consistent for individual operations. Operators describe their practices according to a formatted plan provided by the certification body, which then reviews the plan for supposed compliance with the standard. Such plans, if they appear compliant, are verified in person through an inspection, where deviations from the stated plan are noted along with non-conformities against the standard. The certification body reviews the inspection report and issues a decision, which has several possible outcomes: (i) the operation has irremediable non-conformities and cannot be certified at the present time; (ii) the operation can be certified provided it makes certain corrections ahead of being awarded with certification; (iii) the operation can be certified presently but must take certain corrective actions in due course along a decided timeline; and (iv) the operation is certified with no corrective actions needed. Renewal of the whole process happens annually, although the new EU organic regulation is posing the option of less-than-annual inspection requirements of certain operators based on a risk assessment framework. Inspections may occur at any time and more than once per cycle, dependent on the certification body's determination that such steps are needed to have adequate controls.

Certification carries a certain workload of documentation, time, effort and expense by all parties involved, and in developing economies often proves too costly to be affordable or worth the benefits of the certification itself. In such cases, smallholder producers may be certified as a group, provided they are collectively organized, have more or less similar production styles, market products collectively, and are internally controlled by their own structure. This so-called Internal Control System (ICS) essentially becomes responsible for doing the certification body's work, with the certification body then focusing on the functionality of the ICS itself, especially in terms of assuring all relevant parts of the standard are adequately practised by the groups' member producers and that non-compliant producers' products are detected and effectively removed from the certified stream of commerce. This methodology, i.e. group certification, saves much money and expense for the producers in question but requires a serious undertaking on an ongoing basis to merit certification. The group that is certified and must sell as one entity, or fails as one entity if controls are not adequate. The new EU regulations foresee the option to certify groups within the EU's own borders, which is a departure from the past. At the same time, too strict requirements on the mechanisms employed by ICS could end up becoming too much of a barrier to actually participate in this way.

Organic certification bodies' competency is assured by accreditation bodies, which base their evaluations on ISO Guide 17011: Conformity Assessment – General Requirements for Accreditation Bodies Accrediting Conformity Assessment Bodies. Accreditation bodies may be private or government entities but must be recognized by the scheme owner of the standard/regulation in question. The IFOAM Standard, for example, is accredited by IFOAM's daughter company IOAS (www.ioas.org), whereas in the United States, USDA itself is the accrediting body, and in many other countries, IAF members (International Accreditation Forum) situated in the country in question are the recognized entity (e.g. Inmetro in Brazil, CIQ in China, COFRAC in France, Accredia in Italy, etc.). Not all countries have an IAF member established in country, and in many of these cases, the government either has its own procedures for accreditation. Accreditation to certify organic operations generally entails a specification of scope or categories, of which aquaculture may be considered a separate category of competency, along with crops, livestock or processing. Some accreditations recognize certification of producer groups as a separate category as well.

Other Guarantee Systems

Third-party certification is the most dominant form of organic guarantee in the market, but other methods are growing in their use. Perhaps most notable are Participatory Guarantee Systems (PGS), (https://www.ifoam.bio/en/organic-policy-guarantee/participatory-guarantee-systems-pgs), which are mainly used for more local value chains and rely on horizontal, more intensive participation of producers, consumers and other stakeholders to mutually verify that a standard has been met. PGS afford a greater degree of transparency and familiarity among producers and consumers. Some government regulations formally recognize PGS as an equally valid form of guarantee, most notably Brazil and India, where these types of systems are included in the government regulation and continue to flourish. Other countries' regulations have begun recognizing PGS as well, but in some cases, the approval of such groups to make organic claims is laden with the same kinds of requirements as are imposed on third-party certification, which can be overly burdensome and actually discourages PGS from forming and marketing organic goods; Mexico is an example where this phenomenon has so far occurred (Gould, D., personal communication with Red Mexicana de Tianguis y Mercados Orgánicos 2018).

Direct sale of organic goods by producers to consumers also occurs; the greater the trust between the two, the less necessary an official certification. However, the term "organic", or its recognized legal equivalent in other languages, may be controlled by a national regulation, which may essentially force producers to describe their product in other ways.

Labelling

Products sold as organic generally carry an explicit indication of such, in accordance with the rules in force. The EU regulates not only the term "organic" and its equivalent terms in the different Union languages but also any other terms that suggest that a specific food product has been produced in an organic way. The United States and Japan protect only the term "organic" and its Japanese equivalent respectively.

Some regimes require the use of recognized seals, such as the EU "leaf" logo or the China organic logo, while other regimes make such seal use voluntary, e.g. the USDA seal. Standards have specific requirements for how to use of the organic term, indications of certification, and identification and percentage of organic ingredients in the product. Furthermore, most certifications require some manner of traceability from final product back through the chain of custody. Often this occurs as some kind of lot code or use-by date but can take other forms depending on the case, as determined by the certification body. Requirements for labelling organic goods in the market are also subordinate or otherwise must comply with broader laws pertaining to product labelling.

Import Requirements

The most relevant markets that are importing organic products are from the United States, European Union, Canada and Japan. Other Asian countries such as South Korea and China are fast-growing import markets for organic products. All these countries, together with a growing number of other countries, both in the global North and in the South, have strict rules to access their markets for organic products coming from "third countries". Most of these import regimes recognize the approval/accreditation of the originating product's certification body's respective competent authority, based on compliance or equivalence with the importing regime's legal requirements. Approvals are achieved through either bilateral agreement between countries or direct recognition by the importing country. The United States, EU, Canada and Japan have such bilateral agreements with other countries based on political will and technical assessment that validate the exporting country standard and control system as aligned with that of the importing country. Many of these agreements are reciprocal, such as the one between EU and Canada. In some cases, such as the one between the United States and EU, there are some limitations to the equivalency. This happens by product category, e.g. wines or animal products from EU and pears and apple from the United States, which have additional compliance requirements beyond the equivalency arrangement, or because the whole category is excluded from the agreement. Aquaculture is so far excluded from the scope of these international agreements, with the United States and Japan still without standards on this topic and the Canadian standard only very recently published.

Hints on the Control and Certification System in Europe

To complete the global picture of the control and certification systems, only some elements of the system in force in Europe are mentioned here, which are described in the next chapter. The certification and control systems in the European regulation on organic farming, including organic aquaculture, are complex and differ between member states, which may apply one of the following three types of certification systems: (a) system of private approved inspection bodies; (b) system of designated public inspection authority(ies) and (c) mixed system with designated public inspection authority(ies) and approved private inspection bodies.

Foods may only be marked as "organic" if at least 95% of their agricultural ingredients are organic. The EU organic logo and those of EU member states are used to supplement the labelling and increase the visibility of organic food and beverages for consumers. Products have to bear the name of the producer, the preparer or the vendor and the name or code of the inspection body.

Where the community logo is used, an indication of the place where the agricultural raw materials were farmed must be mentioned, i.e. that the raw materials originate from "EU Agriculture", "non-EU Agriculture" or "EU/non-EU Agriculture".

Organic imports from third countries represent an important part of organic products consumed in most EU member states. This is true also for organic aquaculture products. With the Council Regulation (EC) N° 834/2007 and the Commission Regulation (EC) N° 1235/2008, the framework conditions for imports into the EU changed considerably.

Currently, a system based on the "equivalence" concept has been implemented and is currently widely used. Namely, the EU recognizes imports as equivalent if:

1. The third country in question has been included in the European Commission's list of recognized third countries.
2. The control body issuing the certificate is listed by the European Commission as an "equivalent" control body.

References

Albert J (2010) Innovations in food labelling. FAO and Woodhead publishing limited, Oxford

Arbenz M, Gould D, Stopes C (2015) Organic 3.0 for truly sustainable production and consumption. http://www.ifoam.bio/sites/default/files/organic3.0_web_1.pdf

Arbenz M, Gould D, Stopes C (2016) Organic 3.0 – for truly sustainable farming and consumption. IFOAM Organics International/SOAAN, Bonn

Compagnoni A (2010) Chapter 6: Organic food history and latest trend. In: Albert J (ed) Innovations in food labelling. FAO and Woodhead publishing limited, Oxford

EC (2006a) Council Directive 2006/88/EC Council Directive 2006/88/EC with amendments on animal health requirements for aquaculture animals, and on the prevention and control of certain diseases

EC (2006b) Council Regulation (EC) No 1991/2006 of 21 December 2006 amending Regulation (EEC) No 2092/91 on organic production of agricultural products and indications referring thereto on agricultural products and foodstuffs. Regulation (EC) No 882/2004 of the European Parliament and of the Council of 29 April 2004 on official controls performed to ensure the verification of compliance with feed and food law, animal health and animal welfare rules

EC (2007) Council Regulation (EC) No. 834/2007 on organic production and labelling of organic products and repealing regulation (EEC) No. 2092/91

EC (2008a) Commission Regulation (EC) No. 889/2008 laying down detailed rules for the implementation of Council Regulation (EC) No. 834/2007

EC (2008b) Commission Regulation (EC) No 1235/2008 of 8 December 2008 laying down detailed rules for implementation of Council Regulation (EC) No 834/2007 as regards the arrangements for imports of organic products from third countries

EC (2009) Commission regulation (EC) No. 710/2009 amending Regulation (EC) No. 889/2008 laying down detailed rules for the implementation of Council Regulation (EC) 834/2007, as regards laying down detailed rules on organic aquaculture

EC (2012) Commission implementing regulation (EU) No 508/2012 of 20 June 2012 amending Regulation (EC) NO 1235/2008 laying down detailed rules for implementation of Council Regulation (EC) No 834/2007 as regards the arrangements for imports of organic products from third countries

EC (2013) Commission Implementing Regulation (EU) No. 392/2013 of 29 April 2013, amending Regulation (EC) No. 889/2008 as regards the control system for organic production

Gould D, Kirschenmann F (2006) Tame and wild: organic agriculture and wildness, farming and the fate of wild nature: essays in conservation-based agriculture. Wild Farm Alliance and University of California Press

Huber B, Schmidt O, Batlogg V (2018) Standards and regulations. In: Willer H, Lernoud J (eds) The world of organic agriculture 2018, statistics and emerging trends. FIBL and IFOAM Organics International

IFOAM (2014) Norms for organic production and processing. http://www.ifoam.bio/en/ifoam-standard

IFOAM (2017) General Assembly minutes. https://www.ifoam.bio/sites/default/files/draft_minutes_ga_2017_short_v9_0.pdf

Willer H, Lernoud J (eds) (2017) The world of organic agriculture. Statistics and emerging trends 2017. Research Institute of organic agriculture (FiBL)/IFOAM Organics International, Frick/Bonn. Version 1.1 of February 08, 2017

Willer H, Lernoud J (eds) (2018) The world of organic agriculture. Statistics and emerging trends 2018. Research Institute of organic agriculture (FiBL)/IFOAM Organics International, Frick/Bonn. Version 1.1 February 2018

Chapter 2
EU Regulation on Organic Aquaculture

Emanuele Busacca and Giuseppe Lembo

Introduction

The first EU-wide regulation for the production and labelling of organic products was published in 1991. Council Regulation (EEC) No 2092/1991 was adopted when the European Union, at that time European Economic Community, was composed by 12 Member States.

The first EU Organic Regulation included only rules for plant production and food from ingredients of plant origin and was amended countless times over the following years. The most important addition was the inclusion of production rules for livestock in 1999. Even if, at private level, projects of organic aquaculture were starting in some countries, the new EU rules did not include yet requirements for organic aquaculture.

In 2005 the EU Commission presented a proposal for reviewing the Organic Regulation, and the legislative process begun. After just over 10 years from the first regulation, the organic production and market had grown considerably, the European Union had increased to 25 Member States and the time was ripe for a new organic regulation that would have also allowed the extension of the scope to new categories of products, such as aquaculture products and wine. In 2007 Regulation (EC) No 834/2007 on Organic Production and Labelling of Organic Products was adopted and repealed 1991s Organic Regulation EC (2007).

Such Regulation was lately integrated by Regulation (EC) No 889/2008 laying down detailed rules for its implementation with regard to organic production, labelling

E. Busacca (✉)
IFOAM-EU, Brussels, Belgium
e-mail: emanuele.busacca@ifoam-eu.org

G. Lembo
Stazione Sperimentale per lo Studio delle Risorse del Mare,
COISPA Tecnologia & Ricerca, Bari, Italy
e-mail: lembo@coispa.it

G. Lembo, E. Mente (eds.), *Organic Aquaculture*,
https://doi.org/10.1007/978-3-030-05603-2_2

and control (EC 2008a). The new set of organic rules entered in force on January 1, 2009, although without carrying detailed rules for organic aquaculture. Eventually, in August 2009, Regulation (EC) No 710/2009 laying down detailed rules on organic aquaculture animal and seaweed production was adopted, as amendment to Regulation (EC) No 889/2008, and applied from July 1, 2010 (EC 2009).

In 2009, in addition to the Regulation No 889/2008, there were a number of recognised private standards and two national standards (Denmark and France) for organic aquaculture in Europe.

Several amendments of the organic regulations followed, in order to take into account of the dynamic evolution of the organic sector, the experience gained from the application of these rules and the new technical-scientific knowledge. Then, at the end of 2011, a further wide revision of the organic regulations was announced by the Commission. This was a long-lasting process that reached a first milestone in 2018 when a basic text was approved with the provision that the new regulation will enter into force from January 1, 2021 (EU 2018). This new regulation also provides that further details relating to some defined subjects will be issued with specific delegated acts. However, one new element to mention is the obligation for Member States to establish a free-of-charge public databases to check the availability of organic juveniles at national level. Other indications on how the control system and the import regime will change are described later in this chapter.

Organic Aquaculture in Regulation (EC) No 834/2007

Regulation (EC) No 834/2007 establishes objectives and principles for organic production, labelling, controls and international trade of organic products. The Regulation is composed of 40 recitals, 7 titles with 42 articles and 1 annex.

The recitals are the preambles to the text and set out the reasons for the contents of the articles which follow. The first recital is very important and summarises what organic production is and what it represents for the society: “Organic production is an overall system of farm management and food production that combines best environmental practices, a high level of biodiversity, the preservation of natural resources, the application of high animal welfare standards and a production method in line with the preference of certain consumers for products produced using natural substances and processes. The organic production method thus plays a dual societal role, where it on the one hand provides for a specific market responding to a consumer demand for organic products, and on the other hand delivers public goods contributing to the protection of the environment and animal welfare, as well as to rural development”.

Recital 7 states that “a general Community framework of organic production rules should be established with regard to plant, livestock, and aquaculture production, including rules for the collection of wild plants and seaweeds […]”. For the first time, aquaculture production was included in the EU Organic Regulation.

Recital 17 states that "the implementing rules for livestock production and aquaculture production should at least ensure compliance with the provisions of the European Convention for the Protection of Animals kept for Farming purposes [...]".

Article 1 clarifies that the Regulation applies also to aquaculture, while fishing of wild animals cannot be considered as organic production.

Article 2 sets the definitions for terms used in the Regulation. The definition of aquaculture makes a cross-reference with the definition in Regulation (EC) No 1198/2006 according to which aquaculture is "the rearing or cultivation of aquatic organisms using techniques designed to increase the production of the organisms in question beyond the natural capacity of the environment; the organisms remain the property of a natural or legal person throughout the rearing or culture stage, up to and including harvesting".

In Article 3 there are the general objectives that the organic production should pursue. Among these, it is important to highlight that a sustainable management system should be established that:

(i) *Respects nature's systems and cycles and sustains and enhances the health of soil, water, plants and animals and the balance between them*
(ii) *Contributes to a high level of biological diversity*
(iii) *Makes responsible use of energy and the natural resources, such as water, soil, organic matter and air*
(iv) *Respects high animal welfare standards and in particular meets animals' species-specific behavioural needs*

The regulation also lists, in Article 5, specific principles applicable to farming that organic production should follow. Many of these also apply to organic aquaculture and some are aquaculture-specific:

- Organic aquaculture shall comply with the principle of sustainable exploitation of fisheries.
- Animal health should be maintained by encouraging the natural immunological defence of the animal, as well as the selection of appropriate breeds and husbandry practices.
- The observance of a high level of animal welfare respecting species-specific needs.
- Breeds should be chosen having regard to the capacity of animals to adapt to local conditions, their vitality and their resistance to disease or health problems.
- Animals shall be fed with organic feed composed of agricultural ingredients from organic farming and of natural nonagricultural substances.
- Rearing artificially induced polyploid animals is excluded.
- The biodiversity of natural aquatic ecosystems, the continuing health of the aquatic environment and the quality of surrounding aquatic and terrestrial ecosystems in aquaculture production shall be maintained.

- Aquatic organisms shall be fed with feed from sustainable exploitation of fisheries or with organic feed composed of agricultural ingredients from organic farming and of natural nonagricultural substances.

After stating objectives and principles for organic production, the Reg. (EC) N° 834/2007 establishes basic production rules for the categories of products covered by the scope of the Regulation, that is, plants, seaweeds, animals (including aquaculture species), processed food and processed feed. It also includes the criteria for the products and substances that can be used in those productions such as fertilisers, feed and food additives or products for cleaning and disinfection. Here, in Articles 9 and 10, it is clearly stated that the use of genetically modified organisms "GMOs" and ionising radiation are prohibited in organic production, and this includes organic aquaculture.

According to Article 11, as general principle, the entire holding should be organically managed. However, it is still possible to split the holding into separated production sites, not all managed under organic production, if there is adequate separation between the sites. Furthermore, in case of aquaculture, it is possible to raise the same species both in organic and non-organic. This is an exception that does not apply to terrestrial animals, where the same species cannot be raised both in organic and non-organic in the same holding.

In addition to the general farm production rules laid down in Article 11, basic production rules for seaweed are set in Article 13, while for aquaculture animals are set in Article 15. The main production rules for aquaculture animals are established having regard to:

(a) The origin of the aquaculture animals
(b) The husbandry practices
(c) The breeding
(d) The feed for fish and crustaceans
(e) The bivalve molluscs and other species which are not fed by man but feed on natural plankton
(f) The disease prevention and veterinary treatment
(g) The cleaning and disinfection, products for cleaning and disinfection in ponds, cages, buildings and installations, shall be used only if they have been authorised for use in organic production.

Additionally, the basic requirements for labelling, control and international trade are reported in the Regulation (EC) No 834/2007.

Organic Aquaculture in Regulation (EC) No 889/2008

Regulation (EC) No 889/2008 lays down detailed rules for the implementation of Council Regulation (EC) No 834/2007 on organic production and labelling of organic products with regard to organic production, labelling and control.

This regulation is composed of 5 titles with 97 articles and 20 annexes. It has been amended and integrated by other Commission regulations, eight of them were specifically on aquaculture: Reg. (EC) No 710/2009; Reg. (EU) No 505/2012; Reg. (EU) No 1030/2013; Reg. (EU) No 1364/2013; Reg. (EU) No 1358/2014; Reg. (EU) 2016/673; Reg. (EU) No 2017/838 and Reg. (EU) 2018/1584.

Seaweed Production

Article 6 from (a) to (d) lays down detailed production rules for seaweed. The main provisions regard separation measures, which shall be based on the natural situation, separate water distribution systems, distances, the tidal flow, the upstream and the downstream location of the organic production unit.

Whether it is algae or animal's production, an environmental assessment proportionate to the production unit shall be required for all new operations applying for organic production and producing more than 20 tonnes of aquaculture products per year to ascertain the conditions of the production unit and its immediate environment and likely effects of its operation. The operator shall provide also a sustainable management plan proportionate to the production unit for aquaculture and seaweed harvesting.

Harvesting of wild seaweed shall be carried out in such a way that the amounts harvested do not cause a significant impact on the state of the aquatic environment. Measures shall be taken to ensure that seaweed can regenerate, such as harvest technique, minimum sizes, ages, reproductive cycles or size of remaining seaweed.

Seaweed culture at sea shall only utilise nutrients naturally occurring in the environment or from organic aquaculture animal production. In facilities on land where external nutrient sources are used, the nutrient levels in the effluent water shall be verifiably the same, or lower, than the inflowing water. Only nutrients of plant or mineral origin and as listed in Annex I of the regulation may be used.

Article 6e claims that biofouling organisms shall be removed only by physical means or by hand and where appropriate returned to the sea at a distance from the farm. Moreover, cleaning of equipment and facilities shall be carried out by physical or mechanical measures. In any case, only substances as listed in Annex VII, Section 2 may be used.

Animal Production: General Rules

The rules on animal production in aquaculture are provided in Article 25 from (a) to (t), which is divided into seven sections. The main provisions are set out in the following paragraphs.

This chapter lays down detailed production rules for species of fish, crustaceans, echinoderms and molluscs and applies mutatis mutandis to zooplankton, micro-crustaceans, rotifers, worms and other aquatic feed animals.

Defensive and preventive measures taken against predators under Council Directive 92/43/EEC and national rules shall be recorded in the sustainable management plan.

Farms shall be equipped with either natural filter beds, settlement ponds, biological filters or mechanical filters to collect waste nutrients or use seaweeds and/or animals (e.g. bivalves) which contribute to improving the quality of the effluent.

The competent authority may permit hatcheries and nurseries to rear both organic and non-organic juveniles in the same holding provided there is clear physical separation between the units and a separate water distribution system exists.

Origin of Aquaculture Animal

Locally grown species shall be used, and breeding shall aim to give strains which are more adapted to farming conditions, good health and good utilisation of feed resources.

For breeding purposes or for improving genetic stock and when organic aquaculture animals are not available, wild caught or non-organic aquaculture animals may be brought into a holding. Such animals shall be kept under organic management for at least 3 months before they may be used for breeding.

For on-growing purposes and when organic aquaculture juvenile animals are not available, the introduction into a holding of a quantity, yearly decreasing, of non-organic aquaculture juveniles was allowed by derogation. The derogation expired on December 2016 (EU 2016).

Aquaculture Husbandry Practices

Stocking density and husbandry practices are set out in Annex XIIIa and vary, species by species, from 10 to 25 k/m^3, with the exception of sturgeon for which 30 kg/m^3 is allowed. In considering the effects of stocking density and husbandry practices on the welfare of farmed fish, the condition of the fish (such as fin damage, other injuries, growth rate, behaviour expressed and overall health) and the water quality shall be monitored.

The design and construction of aquatic containment systems shall provide flow rates and physiochemical parameters that safeguard the animals' health and welfare and provide for their behavioural needs. Containment systems shall be designed, located and operated to minimise the risk of escape incidents.

Closed recirculation aquaculture animal production facilities are prohibited, with the exception of hatcheries and nurseries or for the production of species used for organic feed organisms.

Containment systems at sea shall be located where water flow, depth and water-body exchange rates are adequate to minimise the impact on the seabed and the surrounding water body.

Handling of aquaculture animals shall be minimised, and grading operations shall be kept to a minimum and as required to ensure fish welfare.

The use of artificial light shall not exceed 16 h per day, except for reproductive purposes.

The use of oxygen is only permitted for uses linked to animal health requirements and critical periods of production or transport, in the following cases: (a) exceptional cases of temperature rise or drop in atmospheric pressure or accidental pollution; (b) occasional stock management procedures, such as sampling and sorting; and (c) in order to assure the survival of the farm stock.

Slaughter techniques shall render fish immediately unconscious and insensible to pain. Differences in harvesting sizes, species and production sites must be taken into account when considering optimal slaughtering methods.

Breeding

The use of hormones and hormone derivate is prohibited.

Feed for Fish, Crustacean and Echinoderm

Feed for carnivorous aquaculture animals shall be sourced with the following priorities: (a) organic feed products of aquaculture origin, (b) fishmeal and fish oil from organic aquaculture trimmings, (c) fishmeal and fish oil and ingredients of fish origin derived from trimmings of fish already caught for human consumption in sustainable fisheries, (d) organic feed materials of plant or animal origin and (e) feed products derived from whole fish caught in fisheries certified as sustainable under a scheme recognised by the competent authority, in line with the principles laid down in Regulation (EU) No 1380/2013 of the European Parliament and of the Council (EU 2013).

Astaxanthin, derived primarily from organic sources, and histidine produced through fermentation may be used in the feed ration for salmonid.

Where natural feed resources, in ponds and lakes, are not available in sufficient quantities, supplementary feed rations may be allowed, provided that (a) the feed ration of siamese catfish (*Pangasius* spp.) comprise a maximum of 10% fishmeal or fish oil derived from sustainable fisheries and (b) the feed ration of penaeid shrimps comprise a maximum of 25% fishmeal and 10% fish oil derived from sustainable fisheries. Organic cholesterol, preferably, may be used (EU 2014).

In the larval rearing of organic juveniles, conventional phytoplankton and zooplankton may be used as feed (EU 2014).

Specific Rules for Mollusc

Bivalve mollusc farming may be carried out in the same area of water as organic finfish and seaweed farming in a polyculture system.

Provided that there is no significant damage to the environment and if permitted by local legislation, wild seed from outside the boundaries of the production unit can be used in the case of bivalve.

The introduction into a holding of a quantity, yearly decreasing, of seed from non-organic bivalve shellfish hatcheries was allowed by derogation. The derogation expired on December 2016.

Sorting, thinning and stocking density adjustments shall be made according to the biomass and to ensure animal welfare and high product quality. Shellfish may be treated once during the production cycle with a lime solution to control competing fouling organisms.

Bottom cultivation of molluscs is only permitted where no significant environmental impact is caused at the collection and growing sites. The evidence of minimal environmental impact shall be supported by a survey and report on the exploited area.

Disease Prevention and Veterinary Treatments

The animal health management plan in conformity with Article 9 of Directive 2006/88/EC shall detail biosecurity and disease prevention practices (EC 2006a) including a written agreement for health counselling.

The competent authority shall determine whether fallowing is necessary and the appropriate duration which shall be applied and documented after each production cycle.

Ultraviolet light and ozone may be used only in hatcheries and nurseries.

When despite preventive measures a health problem arises, veterinary treatments may be used in the following order of preference: (a) substances from plants, animals or minerals in a homoeopathic dilution; (b) plants and their extracts not having anaesthetic effects and (c) substances such as trace elements, metals, natural immunostimulants or authorised probiotics. The use of allopathic treatments is limited to two courses of treatment per year, with the exception of vaccinations and compulsory eradication schemes. However, in the cases of a production cycle of less than a year, a limit of one allopathic treatment applies.

The withdrawal period for allopathic veterinary treatments and parasite treatments according to paragraph 3 including treatments under compulsory control and eradication schemes shall be twice the legal withdrawal period as referred to in Article 11 of Directive 2001/82/EC.

The Organic Control System

The controls of the compliance with EU food and feed laws are carried out by National Competent Authorities based on Regulation (EC) No 882/2004 (replaced by the Regulation (EU) 2017/625 of the European Parliament and of the Council) (EC 2004; EU 2017).

Those official controls along the agri-food chain apply to both organic and non-organic food and feed marketed in the EU, and they aim to:

- Prevent, eliminate or reduce to an acceptable level risks for the environment, the human beings and the animals
- Guarantee fair practices in the trade of food and feed
- Ensure protection of consumers' interest

In addition to those general controls, organic food and feed and its production, trade and labelling are subject to an additional control system established in the Regulation (EC) No 834/2007 and its implementing regulations. Such control system aims at guaranteeing the production processes and not the products themselves, by verifying and certifying that each operator in the supply chain (farmers, processors, traders, importers) comply with the correct application of the production rules. In practice, it means that the production, trade and marketing of organic products in the EU are subject to a double level of controls: the general official controls carried out to food and feed and the specific organic control system.

According to Regulation (EC) No 834/2007, all the EU Member States have to establish an organic control system (see Fig. 2.1) and designate one or more competent authorities responsible. These are usually the Ministries for Agriculture or the Ministries of Health, but the situation is very different from Member State to Member State.

As a second step, the designed competent authority may delegate the control activities to either a control authority or to one or more control bodies (commonly referred to also as certifiers or certification bodies).

Moreover, in this case, the situation is very diverse in the EU: in most EU countries, the control activities are delegated to private control bodies, while, in few cases, the public control authority stays responsible.

Currently, 6 EU countries left exclusive competence to public authorities (Denmark, Estonia, Finland, Lithuania, Malta, The Netherlands), and 3 EU countries have a mix system with both public authorities and private control bodies

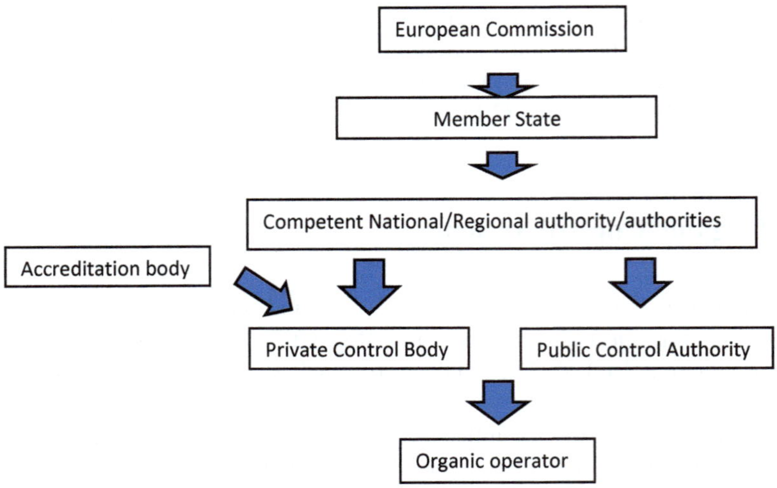

Fig. 2.1 Flow chart of the EU organic control system

(Luxembourg, Poland and Spain), while all the remaining 19 EU countries rely on a system based on delegation to private control bodies.[1]

If the competent authority chooses to delegate control tasks to private control bodies, those control bodies have to be accredited by the appointed National Accreditation Body. To be recognised as control body, many criteria have to be met, e.g. level of expertise, equipment and infrastructure, the qualification and experience of the staff and the impartiality. The competent authority has also to organise audit or inspection to supervise control bodies' activities.

Steps and Obligation for the Organic Operators

All the operators who want to start producing, preparing, storing or importing into the EU organic products, as first step and before placing any organic product on the market, have to notify their activities to the National Competent Authorities and to submit their undertaking to the organic control system (to public control authorities or to private control bodies depending on the EU country's system).

The notification includes the communication of operators' information like the name, the legal and operations' addresses, the nature of operations and products, the undertakings, the date of last application of products and substances non-authorised in organic production and the name of the control body/authority chosen.

The operator has also to draw up and maintain a full description of the unit, the premises and the activity, including the specific characteristics of the production method used. In addition to this, also a description of all the practical measures that are to be taken to ensure compliance with the organic production rules and, very

[1] https://ec.europa.eu/agriculture/organic/consumer-trust/certification-and-confidence/controls-and-inspections/control-system_en

important, a description of the precautionary measures to be taken to reduce the risk of contamination by products and substances that are not authorised in organic production have to be prepared.

Such description has also to include a declaration signed by the operator which include a list of undertakings such as accepting the enforcement of the measures in case of infringement of irregularities or informing in writing the buyers when organic indications are removed from its production.

In the case of aquaculture operators, a full description of the aquaculture installations on land and at sea has to be accompanied, if relevant, by the environmental assessment and by the sustainable management plan of the operation as outlined in points (3) and (4) of Article 6b and, in case of molluscs, in point (2) of Article 25q of Regulation (EC) No 889/2008.

The above-mentioned documents have to be verified by the control authority/body, and, after the verification, the operator has to countersign the verification report and, if necessary, take the corrective measures identified by the control authority/body.

The operator has to keep in the unit or premise all the relevant stock and financial records identifying, e.g. the suppliers, the buyers and the nature and quantity of products delivered to the unit and held in the storage.

According to Article 79b of Regulation (EC) No 889/2008, in the case of aquaculture operators, also a register has to be kept with the following information:

- *The origin, date of arrival and conversion period of animals arriving at the holding*
- *The number of lots, the age, weight and destination of animals leaving the holding*
- *Records of escapes of fish*
- *For fish, the type and quantity of feed and in the case of carp and related species a documentary record of the use additional feed*
- *Veterinary treatments giving details of the purpose, date of application, method of application, type of product and withdrawal period*
- *Disease prevention measures giving details of fallowing, cleaning and water treatment*

The operator has also to grant the control authority/body access to all the facilities, provide any information necessary for the purpose of the control and, if relevant, provide also the results of its own quality assurance system. When the operator also manages units for the production of non-organic animals, also those units will be subject to the controls of control authority/body. The operator has also the responsibility of verifying the organic certificate of the suppliers, called "Documentary Evidence" in the Regulation. Finally, the operator who suspects that a product, either produced by him/her or received by another operator, is not in compliance with organic production rules, has either to cancel any reference to organic from such product or to keep it separated. The operator can only continuing processing or selling that product as organic when the doubt has been eliminated. In case of doubt, the operator has to immediately inform the control body/authority.

The control authority/body has to carry out minimum one physical inspection per year of every organic operator. The number of inspections can be increased to more than one per year according to the results of the evaluation of the risk of the specific operator. Additional random visits are carried out to minimum 10% of the operators, and at least 10% of all the inspections have to be carried out without pre-announcement, so called unannounced inspections.

Basing on the same risk evaluation, every control authority/body has to take and analyse a number of samples corresponding to at least 5% of the total number of operators under its control. Such analyses are needed to detect possible products, substances or production techniques non-authorised in organic production. In addition to that, sample-taking and analyses have to be done any time a suspect arises. In case if production of bivalve molluscs, the inspection visits has to take place before and during maximum biomass production.

When infringements and irregularities are found that affect the organic status of products, the control authority/body applies measures to the operators, according to a catalogue of sanctions adopted by the National Competent Authority.

The control authority/body has to issue a documentary evidence, commonly called "organic certificate" to the operators who meet the requirement of the organic regulations.

The New Organic Control System

Both the general legislation on official controls for food and feed and the organic regulations have been revised over the last year. Regulation (EU) 2017/625 is commonly referred to as the official controls regulation "OCR" and has been adopted in April 2017. It entered into force in the same year and it becomes gradually applicable. The main application date will be December 14, 2019. It is worth pointing out that this new regulation applies to all food and feed, including organic products (EU 2017).

The scope of the new OCR was extended, and now it includes official controls to verify the compliance to food and feed law, animal health and welfare, animal by-products and plant health legislations. The new OCR also clearly includes organics in its scope. Other new aspects of the OCR include the focus on risk-based approach in order to minimise the burden for operators and the reinforcement for financial penalties for frauds and deceptive practices. The OCR has to be further developed and detailed by a high number of Delegated and Implementing Acts, which will mostly be adopted before December 2019.[2]

Considering the two new regulations (New Organic Regulation which will apply from 2021 and the OCR), the main changes for the organic control system relevant to aquaculture production can be identified as follows:

---

[2] https://ec.europa.eu/food/safety/official_controls/legislation_en

- *There is a closer relation with Regulation (EU) 2017/625 on official controls, but additional control rules are described in the new organic regulation.*
- *The documentary evidence will be called "certificate", and there will be an annex showing the model.*
- *Under certain conditions, operators, who sell pre-packed products directly to the final consumer, are excluded from notification and certification obligations.*
- *Member States may exempt operators who sell directly to the final consumer unpacked organic products, other than feed, if these operators sell up to 5,000 Kg of products per year or have an organic turnover not exceeding 20,000 Euro or have a potential certification cost exceeding 2% of the organic turnover. Notification in this case is anyway needed.*
- *Operators for which previous controls have not revealed any non-compliance affecting the integrity of organic products during the last 3 years and operators that are considered as presenting low likelihood of non-compliance can be verified every 24 months (derogation to the mandatory annual inspection).*
- *Group certification will be allowed everywhere in the EU and outside the EU for group of small operators, including the ones producing aquaculture animals. Certain criteria have to be met to be certified under this system.*

Labelling in the European Union

Since 2010, all prepacked organic food and food products, in compliance with the EU Organic Regulation, must be labelled with new EU organic logo (Euro-leaf). When the logo is used, it has to be accompanied by two additional mandatory information:

- The code number of the control body/authority which certifies the product
- The geographical origin of the agricultural raw materials/ingredients of which the product is composed (EU Agriculture or non-EU Agriculture).

The logo is optional for non-prepacked food, organic food imported from outside the EU and organic feed. The logo is forbidden for processed food and feed containing less than 95% organic ingredients, conversion products, products with main ingredients coming from hunting or fishing, and products not (yet) covered by the EU Organic Regulation (e.g. aquaculture species not yet regulated at the EU level). The logo is also forbidden for non-food or feed products, such as seeds for reproduction, ornamentals, cosmetics and textile. According to the new EU organic regulations, which will apply from 2021, the terms "EU/non-EU Aquaculture" will be allowed for indicating the geographical origin of ingredients in the case of aquaculture products.

The Import Regime

While the basic rules are in Regulation (EC) N° 834/2007, the Implementing Regulation (EC) No 1235/2008 lays down detailed rules for imports of organic products from outside the EU (EC 2008). In this regulation, the definition of aquaculture products refers to point 34 of Article 4(1) of Regulation (EU) N° 1380/2013 that is *aquaculture products' means aquatic organisms at any stage of their life cycle resulting from any aquaculture activity or products derived therefrom.*

The organic import regulation covers all kinds of organic products imported into the EU and designates a code to each category of organic products that goes from A to F. The product category "Unprocessed aquaculture products and algae" is designated by the code C.

For organic products produced outside the EU, four different import regimes are provided for (see Table 2.1). However, only two of them are currently applied.

System 1: Recognised Equivalent Third Countries

Non-EU countries whose system of organic production complies with the principles and production rules set out in the EU Organic Regulations and whose control measures are of equivalent effectiveness to those laid down in EU Organic Regulations are eligible for this system. In few words, it means that the EU recognises the organic legislations and the organic control systems of such non-EU countries as equivalent to the EU ones. Currently, specific categories of products from 13 countries are deemed to meet these conditions: Argentina, Australia, Canada, Chile, Costa Rica, India, Israel, Japan, Republic of Korea, Switzerland, Tunisia, the United States and New Zealand. For each country, the regulation specifies which product categories, origin and production standards are accepted, as well as the competent authority and recognised control bodies in that country.

Table 2.1 Import regimes for organic products

Import regime systems		Managed by	Application status
1	List of recognised equivalent third countries	The EU Commission	Yes
2	List of recognised control bodies/authorities for the purpose of equivalence	The EU Commission	Yes, since 2012 onward
3	List of recognised control bodies/authorities for the purpose of compliance	The EU Commission	No, continuously postponed
4	Import authorisations	The EU Member States	Expired in 2012. After receiving an application, Member States carried out an evaluation and released an import authorisation to the EU importers

In the case of aquaculture, unprocessed aquaculture products are not included in any of the above-mentioned recognitions, also because in many cases, organic aquaculture standards are not yet been developed in many non-EU countries.

It means that the import of unprocessed aquaculture products cannot be done through this import system.

System 2: Control Bodies/Authorities Recognised for the Purpose of Equivalence

For non-EU countries that have not yet adopted a national organic legislation or for the ones for which the national organic legislation and/or the organic control system are not considered equivalent to the EU ones, system 2 applies.

This system is based on a list of control bodies and control authorities competent to carry out controls and issue certificates in specific non-EU countries. Those control bodies/authorities can be EU or non-EU based and apply a private standard to certify organic operators operating in non-EU countries. The private standard of every control body/authority is evaluated by the EU Commission and, if deemed equivalent to the EU Organic Regulations, is approved.

Every control body/authority operating outside the EU is recognised for certain non-EU countries and for certain categories of products. All the information related to the recognition of each control body/authority can be found in Annex IV of Regulation (EC) N° 1235/2008. This regulation has been amended several time, therefore it is advisable to refer to the consolidated version.

At the moment, Annex IV includes 57 recognised control bodies/authorities, 13 of them can certify, in specific non-EU countries, operators producing organic unprocessed aquaculture products that can be imported into the EU.

This system came into force on July 1, 2012.

System 3: Control Bodies/Authorities Recognised for the Purpose of Compliance

A third system based on control bodies/authorities recognised for the purpose of compliance was supposed to enter into force in 2011/2012 but has been continuously postponed. Currently it is foreseen to enter into application in 2019.

This approach differs from system 2 only for the fact that the approved control bodies/authorities will have to apply the EU Organic Regulations instead of their own private standards.

System 4: Import Authorisations

Organic food products were also imported on the basis of import authorisations released, by the Member State's competent authority, according to procedures and timing unequal between the various EU Member States, thus creating uneven import conditions. In fact, most organic food products were imported based on such import authorisations.

In 2006 the import regime was changed to simplify procedures by the adoption of Council Regulation (EC) No 1991/2006. The import certificates issued by the competent authorities in the Member States were gradually phased out, and the new import system was based on the establishment of a Third Countries List with equivalent production and inspection procedures, as well as a list of control bodies/authorities competent to carry out inspections and issue certificates in third countries (EC 2006b).

The Organic Import System in the New Organic Regulation

A new EU Organic Regulation will apply from January 1, 2021. Therefore, also the organic import system will be subject to changes. The new import regime will be based on two systems, although a transitional period of validity is foreseen for the current import systems.

The two systems in the organic import system of the New Organic Regulation are (1) trade agreements with Third Countries and (2) control bodies/authorities recognised for the purpose of compliance.

The first system will be similar to the system 1 (see section "System 1: recognised equivalent third countries") applied today, with the difference that all the agreements will be bilateral (the EU recognises as equivalent the non-EU country's organic products only if the non-EU country's recognised the EU organic products, this was not the case today) and all the agreements negotiated by the EU Commission with non-EU countries will have to be confirmed by both the EU Parliament and the EU Council.

The recognition of equivalent countries under the current regime will expire 5 years following the date of application of the new Regulation, and therefore they have all to be renegotiated.

The second system will be very similar to the system 3 (see section "System 3: control bodies/authorities recognised for the purpose of compliance") applied today. The recognition of "equivalent" control bodies under the current regime will expire by up to a maximum of 3 years following the date of application of the new Regulation.

References

EC (2004) Regulation (EC) N° 882/2004 of the European Parliament and of the Council of 29 April 2004 on official controls performed to ensure the verification of compliance with feed and food law, animal health and animal welfare rules

EC (2006a) Council Directive 2006/88/EC with amendments on animal health requirements for aquaculture animals, and on the prevention and control of certain diseases

EC (2006b) Council Regulation (EC) N° 1991/2006 of 21 December 2006 amending Regulation (EEC) N° 2092/91 on organic production of agricultural products and indications referring thereto on agricultural products and foodstuffs

EC (2007) Council Regulation (EC) N° 834/2007 on organic production and labelling of organic products and repealing regulation (EEC) N° 2092/91

EC (2008a) Commission Regulation (EC) N° 889/2008 laying down detailed rules for the implementation of Council Regulation (EC) N° 834/2007

EC (2008b) Commission Regulation (EC) N° 1235/2008 of 8 December 2008 laying down detailed rules for implementation of Council Regulation (EC) N° 834/2007 as regards the arrangements for imports of organic products from third countries

EC (2009) Commission regulation (EC) N° 710/2009 amending Regulation (EC) N° 889/2008 laying down detailed rules for the implementation of Council Regulation (EC) N° 834/2007, as regards laying down detailed rules on organic aquaculture

EU (2013) Regulation (EU) N° 1380/2013 of the European Parliament and of the Council of 11 December 2013 on the Common Fisheries Policy, amending Council Regulations (EC) N° 1954/2003 and (EC) N° 1224/2009 and repealing Council Regulations (EC) N° 2371/2002 and (EC) N° 639/2004 and Council Decision N° 2004/585/EC

EU (2014) Commission Implementing Regulation (EU) N° 1358/2014 of 18 December 2014 amending Regulation (EC) N° 889/2008 laying down detailed rules for the implementation of Council Regulation (EC) N° 834/2007 as regards the origin of organic aquaculture animals, aquaculture husbandry practices, feed for organic aquaculture animals and products and substances allowed for use in organic aquaculture

EU (2016) Commission Implementing Regulation (EU) 2016/673 of 29 April 2016 amending Regulation (EC) N° 889/2008 laying down detailed rules for the implementation of Council Regulation (EC) N° 834/2007 on organic production and labelling of organic products with regard to organic production, labelling and control

EU (2017) Regulation (EU) N° 2017/625 of the European Parliament and of the Council of 15 March 2017

EU (2018) Regulation (EU) N° 2018/848 of the European Parliament and of the Council of 30 May 2018 on organic production and labelling of organic products and repealing Council Regulation (EC) N° 834/2007

Chapter 3
Organic Aquaculture: Economic, Market and Consumer Aspects

Danilo Gambelli, Simona Naspetti, Katrin Zander, and Raffaele Zanoli

Background

The total fish supply around the world has shown a constant increasing trend over the last decades, with growth from less than 20 million metric tonnes in 1950 to more than 169 million metric tonnes in 2015. The main species that are captured are anchovies, Alaska pollock and skipjack tuna, while the main farmed species are finfish (mainly carp) and molluscs (mainly clams). The supply of fish for human consumption was about 142 million tons in 2013 (FAOSTAT 2017). Table 3.1 shows the highly differentiated situation for fish production and the dominant position of the Asian countries in general, where China alone accounts for just over a third of the total world production.

Fish represent an important source of food worldwide as they account for 6.9% of animal protein consumption and 3.5% of total protein consumption (FAOSTAT 2017). Also in this case, there are relevant differences according to geographic area (Fig. 3.1). Despite its high population, China is the country with by far the largest per capita availability of fish, while the availability of fish in Africa and Latin America and the Caribbean is about half of the world's average.

D. Gambelli (✉) · R. Zanoli
Department of Agricultural, Food & Environmental Sciences, Università Politecnica delle Marche, Ancona, Italy
e-mail: d.gambelli@univopm.it; zanoli@agrecon.univpm.it

S. Naspetti
Dipartimento Scienze e Ingegneria della Materia, dell'Ambiente ed Urbanistica, Università Politecnica delle Marche, Ancona, Italy
e-mail: simona@agrecon.univpm.it

K. Zander
Thünen-Institut für Marktanalyse, Thünen Institute of Market Analysis, Braunschweig, Germany
e-mail: katrin.zander@thuenen.de

G. Lembo, E. Mente (eds.), *Organic Aquaculture*,
https://doi.org/10.1007/978-3-030-05603-2_3

Table 3.1 World fish production in 2013 by area [live weight, millions of tons (share, %)]

Measure	World	Africa	North America	Latin America Caribbean	Asia Total	Asia China	Europe	Oceania
Total production of fish	162.8 (100)	9.6 (6)	7.0 (4)	15.0 (9)	113.4 (70)	59.8 (37)	16.3 (10)	1.4 (1)
Non-food fish production	21.4 (100)	0.5 (2)	1.6 (7)	7.0 (33)	9.3 (43)	3.4 (16)	2.6 (12)	0.2 (1)
Total production of fish for food	141.4 (100)	9.1 (6)	5.4 (4)	8.0 (6)	104.1 (74)	56.4 (40)	13.7 (10)	1.2 (1)
Total supply of fish for food (Including import-export)	142.1 (100)	11.2 (8)	7.7 (5)	6.3 (4)	99.8 (70)	52.4 (37)	16.2 (11)	1.0 (1)

Source: Calculated from data from FAOSTAT (2017)

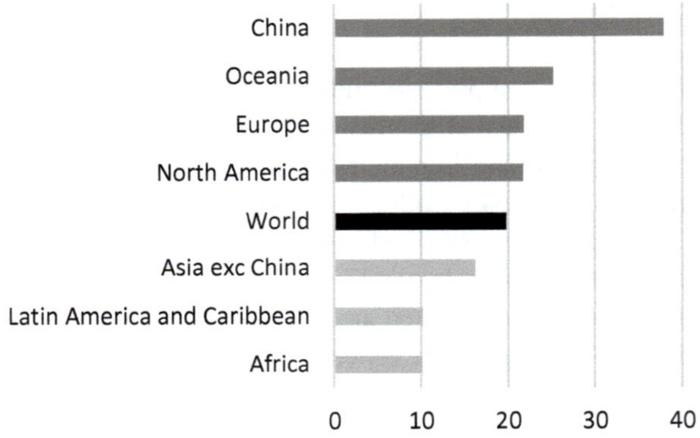

Fig. 3.1 Per capita supply of fish (kg/year) for 2013. (Source: Calculated from data from FAOSTAT (2017))

In this context, the contribution of aquaculture to total fish production has grown constantly over more recent years. According to FAOSTAT (2017), 'the definition of aquaculture is understood to mean the farming of aquatic organisms including fish, molluscs, crustaceans and aquatic plants. Farming implies some form of intervention in the rearing process to enhance production, such as regular stocking, feeding and protection from predators. Farming also implies individual or corporate ownership of the stock being cultivated'. Aquaculture production has been characterised by high, although decreasing, annual growth rates that have ranged from 10.8% in the 1980s to 5.4% for 2010–2015. These growth rates are now declining (although they remain more than double those for captured fish production), and the total production from aquaculture has nearly reached that of captured fish production, accounting for 76.6 million tons and representing 45% of total fish production worldwide (FAOSTAT 2017). By far the largest share of fish production from aquaculture is again China (58%), followed by Indonesia (14.8%). Chinese aquaculture is strongly concentrated on carp farming, which accounts for about 73% of the total freshwater production, while molluscs account for 78% of the marine production (FAO 2014). Asian countries in general are among the most important producers of

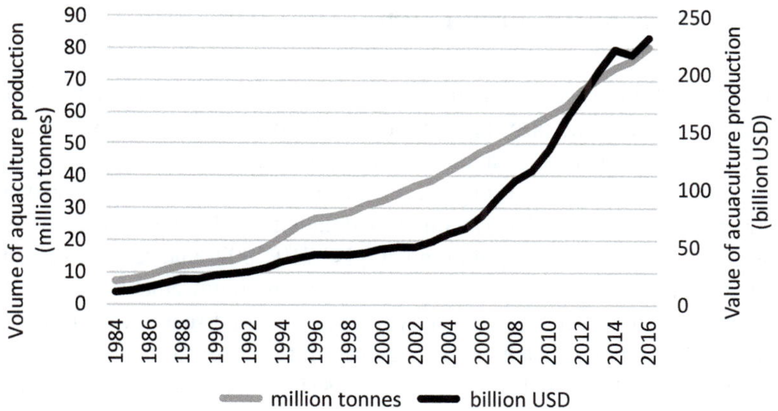

Fig. 3.2 Volumes and value of fish production from aquaculture worldwide. (Source: Calculated from data from FAO – Fisheries and Aquaculture Department – Global Production Statistics 2018)

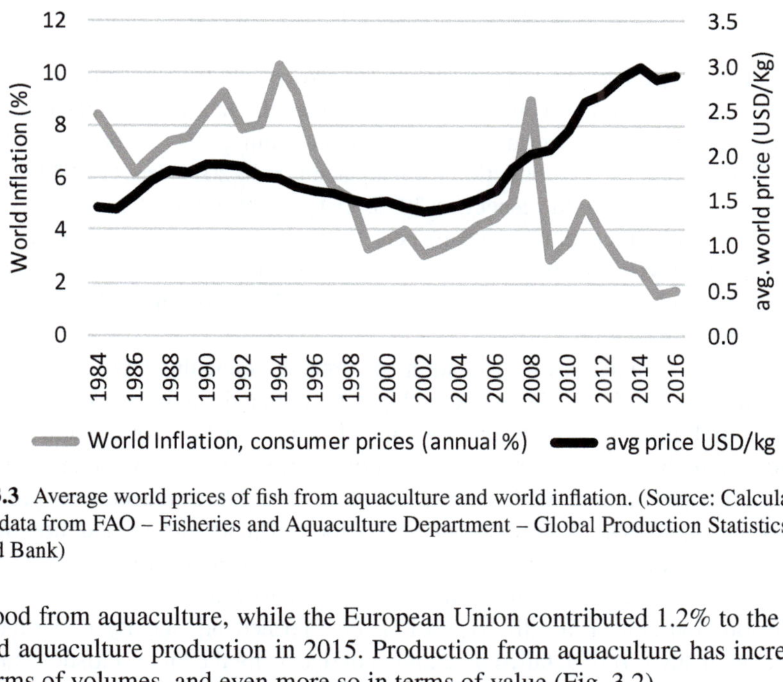

Fig. 3.3 Average world prices of fish from aquaculture and world inflation. (Source: Calculations from data from FAO – Fisheries and Aquaculture Department – Global Production Statistics, and World Bank)

seafood from aquaculture, while the European Union contributed 1.2% to the total world aquaculture production in 2015. Production from aquaculture has increased in terms of volumes, and even more so in terms of value (Fig. 3.2).

The average world price of aquaculture products for farmers ranged from 1.4 to 2.0 USD/kg for about 20 years and then showed substantial growth from 2008. The price trend does not appear to be conditioned by any particular relationship to inflation and might instead be due to growth in the demand for aquaculture products (Fig. 3.3).

Organic Aquaculture Production

The adoption of an organic system for aquaculture requires compliance with general principles of organic farming, which include specific regulations and certification schemes. The general standards for organic aquaculture were defined by IFOAM (2006). Provisions for organic standards for aquaculture were also defined in Regulation (EC) N° 834/2007, with more detailed regulations in Regulation (EC) N° 889/2008 (amended by Regulation [EC] N° 710/2009) and respective Annexes. European Union regulations define the guidelines and requirements on fish origins, husbandry, breeding, feeding, veterinary treatments and disease prevention.

About 80 different private organic aquaculture standards have been defined, many of which relate to the European countries (Prein et al. 2012). A description of the evolution of certification standards for organic aquaculture is available in Bergleiter et al. (2009). Although organic standards share common principles, some differences are however encountered (Mente et al. 2011). The main aspects to consider at the farm level relate to the conditions for the aquatic environment, breeding, nutrition, husbandry practices (e.g. stocking density requirements) and animal welfare (e.g. veterinary treatments).

Production from organic aquaculture has grown rapidly over recent years but remains at relatively low volumes at the world level. Data on volumes of organic aquaculture production worldwide are available from 2017 (Lernoud and Willer 2017 2018). Total world production from aquaculture in 2016 was 415,554 mt, with an increase of 8.2% with respect to 2015. Despite the relevant growth rates, the share of organic aquaculture with respect to total aquaculture remains at around 0.5%. It is necessary to specify, however, that data on organic aquaculture are still very sparse and are missing for many countries. Therefore, these data need to be interpreted with caution.

Figure 3.4 summarises the situation for organic aquaculture production in the world in 2015–2016 and shows the highly differentiated situation both in terms of volumes and growth rates. Despite the limited share for China for organic aquaculture (0.5% in 2015), it still maintains the dominant position for volume of organic aquaculture production, with over 74% of total organic aquaculture products. European countries account for about 20% of the world organic aquaculture. Ireland is the most relevant producer in Europe, with 40,873 metric tonnes produced in 2016, and with an annual growth rate of 31% from 2015 to 2016.

Information concerning the type of species farmed organically is particularly scarce, with any sort of detailed breakdown of data from the main statistical sources only available for 17% of the total organic aquaculture production. These data also refer mainly to the European countries. Details concerning the relative weights for organic aquaculture in China are available from Xie et al. (2013). We have used these data to provide an estimate of the organic aquaculture volume per species in 2016 (Fig. 3.5).

Detailed information of the economic aspects of organic aquaculture in Europe is available from EUMOFA (2017), and from reports of the OrAqua project

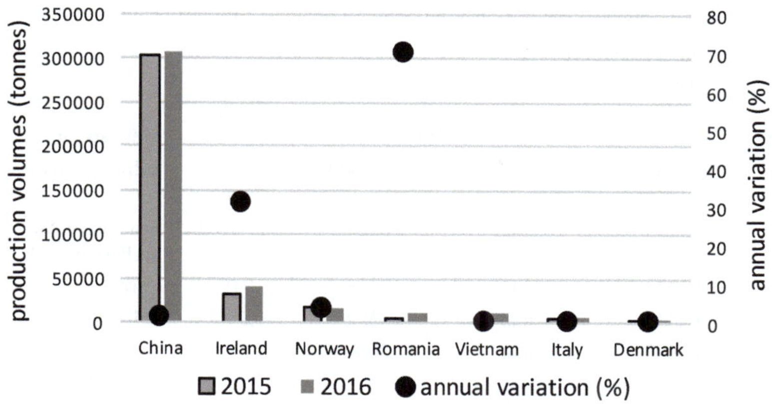

Fig. 3.4 Organic aquaculture production, as the main producers* and the annual variation. (Source: Calculated from FiBL data; *95% of total organic world production; Vietnam: 2015 not available)

Fig. 3.5 Estimated organic aquaculture production by species in 2016. (Source: Calculated from Lernaud and Willer (2018) and Xie et al. (2013))

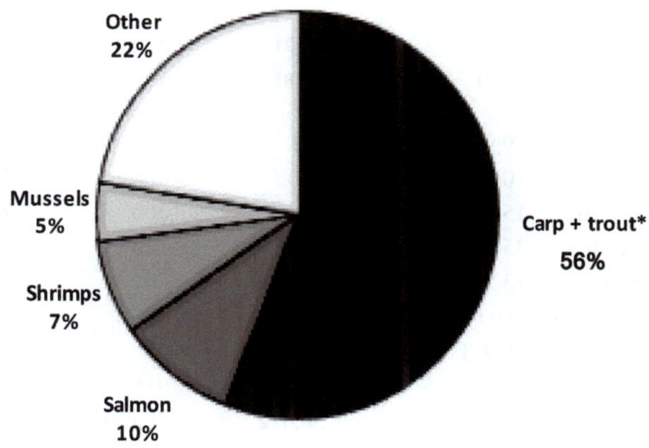

(www.oraqua.eu) for an economic analysis (Prins et al. 2015). Despite the general positive trend for organic aquaculture in Europe, production is not evenly distributed, and some countries have even reduced their production over recent years, particular for Belgium, Croatia, Germany and the UK. The share of organic aquaculture is very variable across the European countries, with Ireland and eastern European countries showing the highest values (Table 3.2). Surveyed data from EUMOFA and EUROSTAT in some cases are different, with the latter showing in general higher values for organic aquaculture volumes. The main organically farmed species in Europe in terms of volumes are salmon (strong concentration of production in Ireland), mussels (mainly produced in Italy and Ireland), trout (mainly produced in France and Denmark), carp (mainly produced in Hungary, Poland and Romania), sea bass and sea bream (mainly produced in Italy and Greece) (Fig. 3.6).

Table 3.2 Organic aquaculture: volumes and share of the main European Union producers (2015)

	Total aquaculture (tons)	Organic aquaculture (tons)			
Country	(FAO)	(EUROSTAT)	(EUMOFA)	Share (%)	Main species
Ireland	39,650	31,227	22,000	55.5–78.9	Salmon, mussels
Italy	148,763	5492	8500	3.7–5.7	Sea bass/bream, trout, mullet, mussels
Hungary	17,337	3498	3498	20.0	Carp
UK	206,834	n.a.	3382	1.6	Salmon
France	206,800	n.a.	3000	1.5	Salmon, trout, sea bass/bream, mussels
Denmark	35,867	2934	2864	8.0–8.2	Trout, mussels, sea bass/bream
Romania	11,042	6384	2042*	18.5–57.8	Carp
Spain	289,821	2709	1353	0.5–0.9	Trout, mussels, sea bass/bream
Portugal	9322	1300	1300	13.9	Mussels
Lithuania	4450	1300	1117	25.1–29.2	Carp
Germany	29,909	621	621	2.1	Carp, trout
Greece	106,118	720	400	0.4–0.7	Sea bass/bream
Croatia	15,572	300	300	1.9	Sea bass, mussels
Austria	3503	n.a.	120	3.4	Carp
Bulgaria	13,537	80	80	0.6	Mussels
Slovenia	1607	32	32	2.0	Mussels
Poland	36,971	18	19	0	Carp, trout
Latvia	863	7	9**	0.8–1.0	Carp
Total	**1,177,966**	**56,622**	**50,637**	**3.9–4.4**	

Source: Calculated from EUMOFA (2017); *2014; ** 2016

The Economics of Organic Aquaculture

Conversion to organic aquaculture is a complex process that involves a multidimensional approach that covers social, economic and environmental issues (Bellon and Lamine, 2009). Any decision to convert is of course influenced by context-specific issues that can be extremely different for different countries. Following Padel (2001) and Stofferahn (2009), we can argue that reasons to convert to organic aquaculture might be mainly classified as farming aspects, such as technical and production issues, economic and financial evaluations, and personal motivation of farmers, such as personal health. Motivation to convert to organic aquaculture in Asian countries, for instance, might include the need for alternative trade opportunities that are more oriented towards sustainability and social inclusion (Omoto and Scott 2016; Ahmed et al. 2018) or the possibility to exploit economically efficient integration

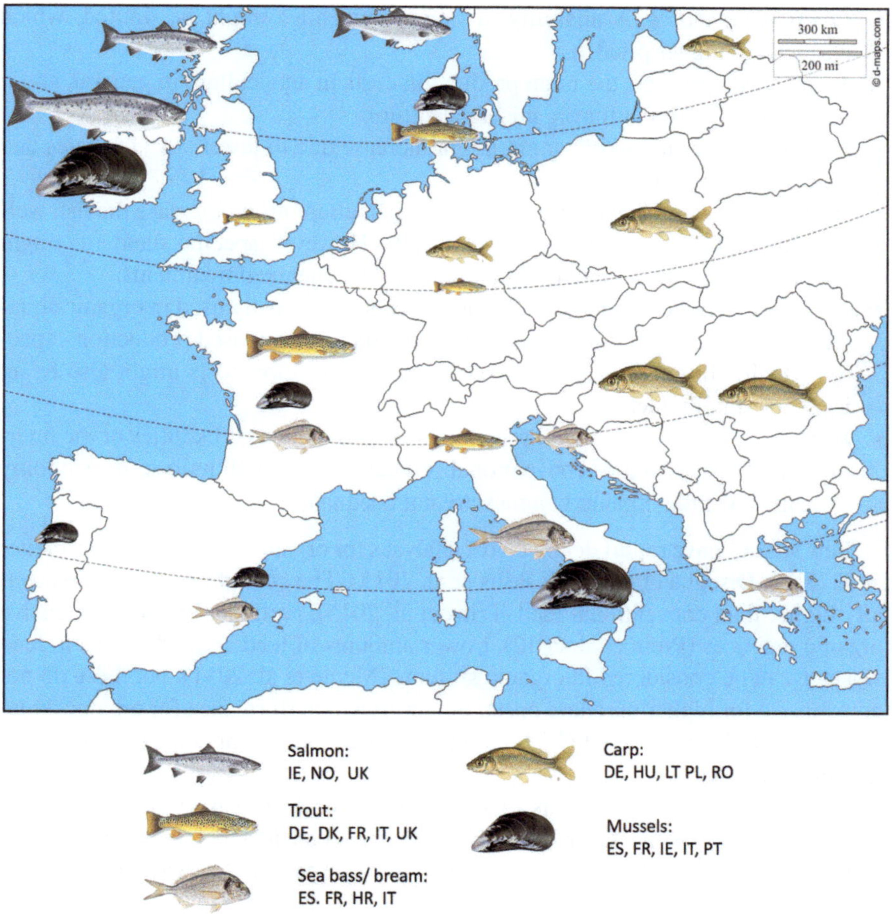

Fig. 3.6 Main organic aquaculture producers in Europe (icon sizes are approximatively proportional to volumes of production). (Source: Adapted from Prins et al. (2015) and EUMOFA (2017))

with other farming production (Nair et al. 2014). Organic aquaculture is considered an opportunity for rural development and poverty reduction in developing countries (Prein et al. 2012). In the western context, obstacles to convert to organic aquaculture might also depend on the perception of organic farming practices as not oriented to efficient production systems (Home et al. 2018). For the particular economic motivations to convert to organic aquaculture production, the main aspects to take into consideration are those of production at the farm level, the processing and marketing and market conditions and demand and consumer attitudes.

The requirements for organic aquaculture have specific consequences according to the following economic aspects:

- *Stocking density*: Organic standards might require reduced stocking rates (see, e.g. Naturland 2017), which will result in higher average fixed costs per unit of output.

- *Livestock/juveniles*: Availability of organic juveniles might be limited, which will lead to higher prices.
- *Feed*: This represents the main production costs in aquaculture in general, and it is particularly critical for organic aquaculture.
- *Labour*: Labour units per output might increase due to lower stocking rates and longer growth periods.
- *Welfare*: The extensive nature of organic aquaculture might increase the fish welfare and reduce the necessity for treatments. However, specific standards might significantly constrain the use of antibiotics and chemical treatments.
- *Investments and general costs*: Ponds and cages are mostly used in organic aquaculture. A stocking density constraint might require increased production capacity, which will result in higher fixed costs. Certification costs might also be an issue for smallholders.
- *Processing and distribution*: Processing requires dedicated facilities or the interruption of processing of conventional products, which will reduce the economy of scale when the organic volumes are not adequate.

Few studies have report detailed cost analyses for organic aquaculture (Bergleiter et al. 2009; Disegna et al. 2009; Prins et al. 2015). The cost of organic feed is usually higher than conventional feed (Prins et al. 2015), particularly due to extensive organic practices (Prein et al. 2012). Lower amounts of feed and reductions in feed wastage can be considered in organic systems (Mente et al. 2011), but these do not compensate for higher purchase costs. Better growth performance for organic aquaculture might be considered (Di Marco et al. 2017) and might contribute to reduction in feed costs.

Figure 3.7 shows a synthesis of the main results from case studies regarding production costs at the farm level. The countries considered in the analysis are among the main producers in the respective fish species: Norway, the UK and

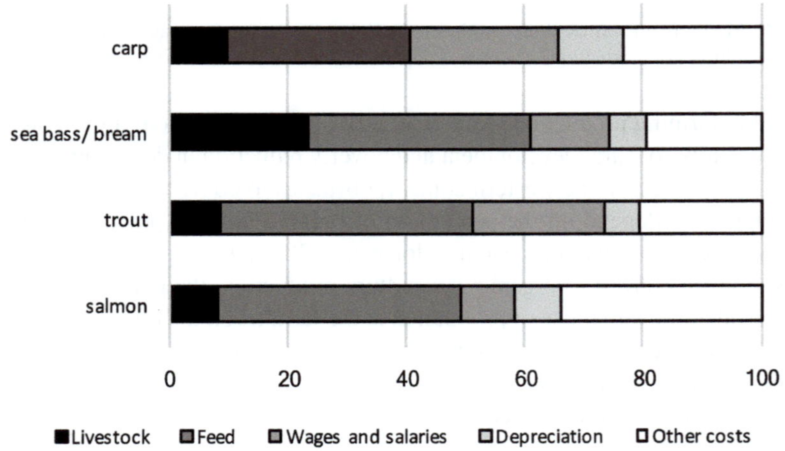

Fig. 3.7 Cost-price structure for selected organically farmed species in selected European countries (average values, % of total costs) (Source: Calculated from Prins et al. (2015))

Ireland for salmon; Denmark, France and Italy for trout; France, Italy and Spain for sea bass and sea bream; and Romania, Poland and Germany for carp. Simple arithmetic averages across the countries have been calculated to summarise the data on the cost share. While such averages might be conditioned by specific conditions for an individual country, these results provide an initial insight into the value chain of organic aquaculture. Although differences in the cost structures are evident across species, the feed share is predominant in all cases. Other costs (e.g. general costs, energy, maintenance, financial costs) and labour costs have relevant but differentiated importance according to the type of species, while the livestock (e.g. cost of juveniles) is particularly relevant for sea bass/bream.

Table 3.3 shows a comparison of the costs and other relevant economic factors for organic aquaculture with respect to conventional aquaculture. The increase in the general costs is particularly disadvantageous for organic salmon production, given the need for larger facilities as a consequence of reduced stocking rates. Similar or lower requirements for feed are observed for all species. The reduction in the quantity is, however, more than compensated for by the price differences, particularly in the case of organic carp. Given the relevance of the share of feed to total costs (Fig. 3.7), 65% of the total difference in cost for organic carp farming with respect to conventional systems is a result of the feed costs. About 35% of the higher production costs for organic sea bass/bream are due to the price difference for organic juveniles (Prins et al. 2015). The purchase prices for organic feed and juveniles are also relevant for sea bass/bream farming. An extra cost of 60% for sea bass

Table 3.3 Main productivity and cost categories for selected species in Europe[a]: relative changes (%) of organic vs conventional aquaculture where not differently specified

Factor	Measure	Change in costs from conventional to organic aquaculture (%)			
		Salmon	Trout	Sea bass/bream	Carp
Stocking density		−40	−15	−15	=
Daily growth		−35	=	−20	−10
Feed	Quantity	−15	=	=	−10
	Price	+12.5	+30	+50	+100
Livestock	Quantity	=	=	=	+50 to +100
	Price	=	=	+50	=
Labour		+15	+15	+15	+10
Mortality rates		=	=	=	=
Health costs		=	=	=	=
Other costs[b]		+150	=	=	+7
Certification		(€3000/year)	(€600/year)	(€600/year)	(€600/year)
Overall cost difference		**+23 to +40**	**+15 to +18**	**+29 to +42**	**+31 to +81**

Source: Prins et al. (2015)

[a]Salmon, Norway, the UK, Ireland; trout, Denmark, France, Italy; sea bass/bream, France, Italy, Spain; carp, Romania, Poland, Germany

[b]Depreciation, maintenance, financial costs

= no difference

organic farming was also reported by EUMOFA (2017). The combination of these factors leads to a generalised increase in the production costs for organic aquaculture with respect to conventional systems, which will normally range between 15% and 81% across the different species and countries.

The cost differences reported in Prins et al. (2015) are lower for trout production, while they are particularly high for organic carp production in Romania. For organic trout farming in Italy, Disegna et al. (2009) consider 20–30% cost increase compared to conventional aquaculture, which was mainly due to higher costs for feed and monitoring, and to an increase in the unitary fixed and general costs due to lower stocking density. Higher cost differences (35%) were reported by EUMOFA (2017) for a case study in France.

While the general consensus can be considered as higher production costs for organic aquaculture, a more differentiated picture emerges in terms of yields. Prins et al. (2015) showed a generally negative situation for productivity of organic aquaculture, due mainly to reductions in the stocking rate, which is particularly relevant for salmon, and the daily growth, which might be related to lower feed conversion rates. Stocking rates are 40% lower for organic salmon farming, and 15% lower for trout and sea bass/bream. The reductions in daily growth have been reported as 35% for salmon, 20% for sea bass/bream and 10% for carp, with no significant reduction for trout (Prins et al. 2015). Disegna et al. (2009) also considered yield reductions for organic trout farming in Italy due to the lower stocking rates. However, Di Marco et al. (2017) carried out a comparison of organic and conventional farming of sea bass in Italy and indicated good performance for the organic system. In particular, the growth of organic sea bass was more rapid, mainly due to their higher feed intake, an improved protein/fat ratio and higher protein availability in the organic feed. For prawn and shrimp farming, organic systems appear to obtain higher yields than conventional farming. Paul and Vogl (2012) reported comparatively higher average yields for organic shrimp farming in Bangladesh, particularly for small farms where animal welfare is better due to more frequent water exchange and better quality of feed. Reported yield improvements ranged from 80 to 260 kg/year/ha in Bangladesh and India. Nair et al. (2014) analysed an integrated rice-prawn system in India that provided lower rice yields, which were compensated for by 10% higher prawn yields.

The general picture that is emerging from this analysis of the production process shows higher costs and a differentiated situation in terms of productivity in organic aquaculture. This implies that overall profitability is strongly conditioned by the higher prices obtained. These premium prices at the farm level are positive and range from 20% to 200%, depending on the farmed species and the country (Table 3.4). Higher prices for farmers are reported for China, which is particularly relevant given that Chinese production of organic trout and carp is the largest around the world. Conversely, price premiums for European countries are generally lower and are not always sufficient to ensure adequate profitability for the organic sector.

Prins et al. (2015) showed a critical situation for carp and sea bass/bream farming in particular. For some countries (e.g. Poland for carp; Italy, Spain for sea bass/bream), total costs exceeded the farm gate prices, which resulted in negative margins. Where available (e.g. France for sea bass/bream; Germany, Romania for carp),

Table 3.4 Reported relative increases in prices obtained and profitability from organic aquaculture as compared to conventional aquaculture

Factor	Source	Country/ area	Increased price obtained/profitability from conventional to organic aquaculture (%)						
			Salmon	Trout	Sea bass/ bream	Carp	Mussels	Shrimp/ prawns	Seafood/ others
Increased price obtained (farm level)	Ankamah-Yeboah et al. (2017)	Denmark		33					
	EUMOFA (2017)	Europe[a]		73			20		
	Prins et al. (2015)	Europe[b]	30	30	35	30			
	Xie et al. (2013)	China		75 to 117		75 to 117		50 to 87	18 to 200
Increased price obtained (consumer level)	Ankamah-Yeboah et al. (2016)	Denmark	20						
	Asche et al. (2013)	UK	25 (frozen fillet)						
	Disegna et al. (2009)	Italy		46					
	EUMOFA (2017)	Europe[a]		20 to 25			20 to 25		
	Olesen et al. (2010)	Norway	15						
	Paul and Vogl (2012)	Bangladesh						14 to 20	
	Prins et al. (2015)	Europe[b]	40 to 50 (smoked/frozen)	50 to 60 (whole)	50 (frozen/ whole)	50 (frozen/ whole)	13 to 25		
Margin/profitability (farm level)	Ankamah-Yeboah et al. (2017)	Denmark		−6 to 8[c]					
	Prins et al. (2015)	Europe[b]	0 to 40[d]	10 to 35[d]	−12 to 9[d]	−1 to 31[d]			
Margin difference (farm level)[e]	Nair et al. (2014)							117	
	Prins et al. (2015)	Europe[b]	−9 to 8[d]	12 to 13[d]	−8 to 1[d]	−52 to −1[d]			

[a]Trout, France; mussels, France, Bulgaria (consumer price only)
[b]Salmon, Norway, the UK, Ireland; trout, Denmark, France, Italy; sea bass/bream, France, Italy, Spain; carp, Romania, Poland, Germany
[c]Net return on asset
[d]Margin on cost price
[e]Calculated from Prins et al. (2015)

positive margins are however entirely dependent on the subsidies that are available, which represent the main condition for farm profitability. While conventional carp farming provides adequate profitability, the premium prices for organic aquaculture are still not sufficient to compensate for the extra costs.

Salmon and trout production are characterised by better economic performances. Here, the margins are positive for all species, even without considering the subsidies, with the exception of an Irish case study that showed a slightly negative margin, excluding subsidies.

Comparing the profitability of organic aquaculture with the conventional system, the differences in the relative margins relating to total costs (including subsidies) for organic and conventional systems are shown in Table 3.4. Conversion to organic production is particularly critical for carp, while the differences for sea brass/bream margins are more limited, also due to the scarce profitability reported in the conventional case studies considered for these comparisons. For salmon, the comparisons of the profitability indicate a better performance for UK organic farming only and negative differentials particularly for the Irish case. Organic trout farming provides the best performance if compared to the conventional aquaculture. These data were confirmed by Ankamah-Yeboah et al. (2017), who analysed the profitability of organic farming of trout in Denmark by measuring the rate of profitability based on the net return (operational profit, i.e. owner remuneration) on assets over 3 years. For 1 of the 3 years considered, they reported negative results for profitability, which was attributed to the specific inclusion of investment costs. A comparison with the profitability of conventional aquaculture, and for conventional dairy and agriculture farming, showed a globally positive performance for organic trout farming.

Nair et al. (2014) showed some considerable increases in profitability for organic farming, such as the combination of rice and prawn organic production. This system is particularly effective for prawns, where the net margin increased by 117% with respect to the conventional system.

For the distribution costs, Figs. 3.8, 3.9, 3.10 and 3.11 show the make-up of price, i.e. unitary revenue, along the main stages of the supply chain in relative terms. These data are based on Prins et al. (2015), based on a calculation model including production and processing costs, yields and losses, revenues for nonedible parts and sale prices. As a general consideration, the component of price obtained by farmers is always lower in relative terms for organic aquaculture, although this is counterbalanced by higher gains at the distribution level. Conversely, the costs of gutting and fillet processing are higher for the conventional system. In absolute terms, the distribution margins are highest for salmon, which is considered a luxury product, and particularly as smoked salmon. Indeed, the distribution margins for smoked salmon are the highest both in relative and absolute terms, while they are the lowest for carp fillets.

The highest processing and distribution costs for all organic cases provides an indication of the effects of diseconomies of scale, due to the limited market size for organic aquaculture. Higher costs at the processing and distribution levels are the main source for the premium consumer prices for organic aquaculture, which appear not to be distributed proportionally to the farmers.

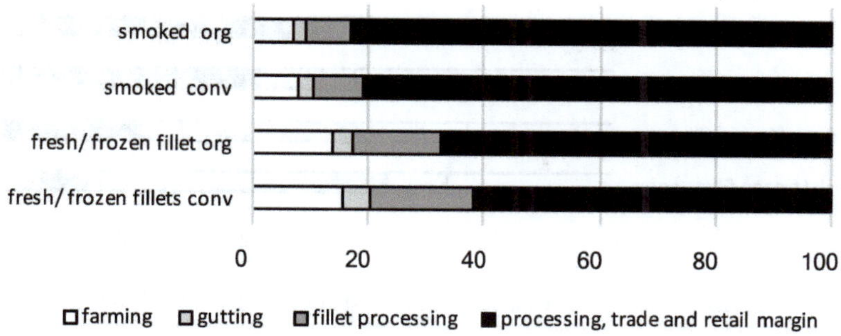

Fig. 3.8 Make-up of price for farmed fish along the supply chain (% of total unitary price at consumer level): salmon. (Source: Calculated from (Prins et al. 2015))

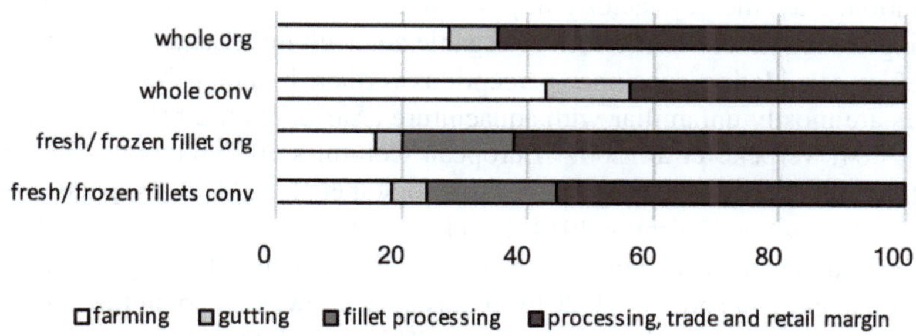

Fig. 3.9 Make-up of price for farmed fish along the supply chain (% of total unitary price at consumer level): trout. (Source: Calculated from (Prins et al. 2015))

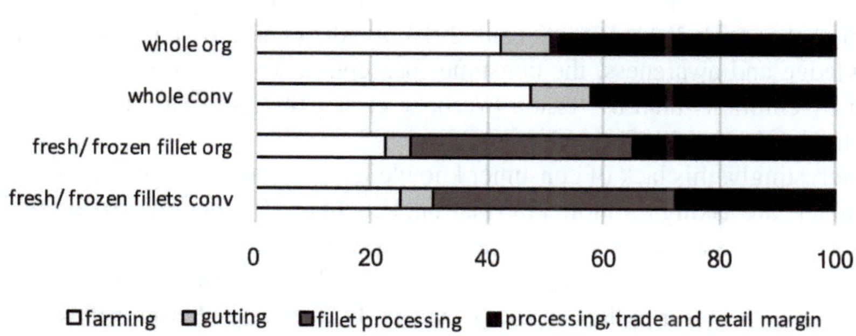

Fig. 3.10 Make-up of price for farmed fish along the supply chain (% of total unitary price at consumer level): sea bass/bream. (Source: Calculated from (Prins et al. 2015))

Consumer Awareness and Product Knowledge

Products from organic aquaculture have only recently gained importance in the market. This is why little attention has been directed so far towards consumer knowledge and perception of organic aquaculture (Schlag and Ystgaard 2013). However, in the policy actions to define organic aquaculture, consumer perception

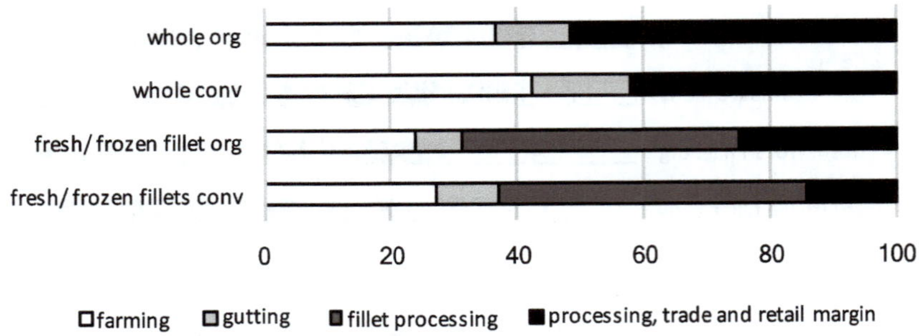

Fig. 3.11 Make-up of price for farmed fish along the supply chain (% of total unitary price at consumer level): carp. (Source: Calculated from (Prins et al. 2015))

is mentioned as 'the key factor that policy makers should take into account when defining the regulatory framework of organic aquaculture' (Lembo et al. 2018).

Before considering consumer perceptions further, it should be noted that consumers are mostly unfamiliar with aquaculture (Aarset et al. 2004; Arvanitoyannis et al. 2004; Verbeke et al. 2007; European Commission Directorate-General for Maritime Affairs and Fisheries 2009; Altintzoglou et al. 2011; Schlag and Ystgaard 2013; Gutierrez and Thornton 2014; Feucht and Zander 2015; Zander et al. 2018). Many consumers are even not able to distinguish between wild-caught fish and fish from aquaculture (Zander et al. 2018). Consumer knowledge about the sustainability in aquaculture production is even lower. This limited knowledge can result in a 'romanticised' and misleading image of aquaculture, and it gives room for the potential to influence consumer ideas and concerns about intensive terrestrial animal husbandry (Honkanen and Ottar Olsen 2009; Stubbe Solgaard and Yang 2011; Vanhonacker et al. 2011; Pieniak et al. 2013; Zander et al. 2018). Due to this lack of knowledge and awareness, the consumer perception of aquaculture can be driven more by emotions than by reason (Verbeke et al. 2007; Vanhonacker et al. 2011; Feucht and Zander 2015).

Interestingly, this lack of consumer knowledge does not automatically mean that consumers are asking for more information. Feucht and Zander (2015) reported that some consumers are aware of their lack of knowledge and want more transparency and information concerning aquaculture in general. They ask for standardised and comprehensible information on the packages. In contrast, other consumers do not wish to know more about the fish farming because they fear that more information might be confusing, as they already have information overload; it might even cause them to stop consuming these organic aquaculture fish altogether.

Labels and Label Knowledge

Labels are an important means for communicating the various attributes of a product to the consumer. These are of particular relevance in the case of 'credence goods', where the specific characteristics cannot be verified during or after purchase and even during consumption. Several labels exist in European markets that indicate the use of sustainable aquaculture practices. These include the Aquaculture Stewardship Council (ASC), the Friends of the Sea (FOS), and also European and national organic labels. The low consumer knowledge of (sustainable) aquaculture also corresponds to low knowledge of sustainability labels in general. Consumer confusion is generally enhanced by the proliferation of too many 'eco-labels' (Langer et al. 2007), and this also applies to organic aquaculture products (Altintzoglou et al. 2010; Feucht and Zander 2015; EUMOFA 2017). However, consumer knowledge about organic labels is relatively high, as these are used on all food products, and not only on farmed fish, like for the ASC and FOS labels (Grunert et al. 2014; Zander et al. 2015, 2018; Ankamah-Yeboah et al. 2016).

Similar data have been reported relating to consumer trust: organic labels are trusted more compared with other sustainability labelling of products, although some consumers have low confidence in the certification process and compliance with respect to organic standards (Feucht and Zander 2015; Ankamah-Yeboah et al. 2016). Indeed, trust in a logo is an important mediator between labels and consumer choice (Zanoli et al. 2015). Distrust of organic labels and certification has, however, been reported for all categories of organic products, especially by the occasional consumer (Zanoli and Naspetti 2002; Zander et al. 2015). According to Feucht and Zander (2015), the consumer who is interested in sustainability issues appears to prefer organic aquaculture products and products from sustainable wild fisheries, even though a general lack of label knowledge has been reported.

According to EUMOFA (2017), organic aquaculture needs to be clearly differentiated from other competing schemes, such as eco-labelled or 'sustainable' aquaculture. Therefore, the 'credibility and readability of organic labels in front of eco-labels' should be increased by improved communication, with stressing of the 'high-level principles of sustainability and animal welfare (in addition to food quality objectives)' (EUMOFA 2017).

Consumer Attitudes

As a consequence of the relative lack of knowledge, many consumers do not clearly distinguish between sustainable and organic aquaculture (Feucht and Zander 2015; EUMOFA 2017). Indeed, the two terms are frequently mixed or used synonymously. Sustainable aquaculture is expected to avoid drug use as far as possible and to work without artificial additives and hormones (Stubbe Solgaard and Yang 2011; Kalshoven and Meijboom 2013; Almeida et al. 2015; Feucht and Zander 2015; Zander et al. 2018). Consumers believe that sustainable aquaculture should be a 'natural' way of production that respects the fish welfare and the environment (Schlag and Ystgaard 2013; Feucht and Zander 2015). 'Mass production' is not perceived as sustainable, and fish feed should also be sustainable and species-appropriate (Zander et al. 2018). Moreover, full transparency along the supply chain and outstanding quality are demanded by the consumer. This would imply greater collaboration along the supply chain (Naspetti et al. 2017), although organic systems do not appear to perform better than other systems in this respect (Naspetti et al. 2011). With regard to closed recirculation systems, there are some associations with 'mass animal husbandry'. The welfare of the fish is not generally believed in these systems, and the ecological advantages with respect to nutrient run-offs are frequently outweighed by the lack of naturalness and the assumed deficiencies towards fish welfare (Feucht and Zander 2015). A lack of product knowledge might also be the reason for the unclear attitudes of the consumer (EUMOFA 2017).

Despite a general lack of knowledge, some consumers have relatively clear conceptions and expectations of organic aquaculture. Whereas sustainability is a more or less vague term with an unclear definition for most consumers, on the other hand, organic is perceived as a fixed term that is familiar to many consumers. Some consumers (mainly those who regard themselves as organic consumers) know that there is a regulatory framework that defines organic aquaculture (Feucht and Zander 2015). Those who know about organic aquaculture perceive it to be the ideal aquaculture practice, and they appreciate seafood from organic aquaculture. They argue that all sustainable aquaculture should follow organic standards in order to avoid misunderstandings (Feucht and Zander, 2015; Risius et al. 2017; Zander et al. 2018).

Consumer perceptions of what the product attributes are that make farmed fish organic aquaculture are mostly in line with current organic aquaculture practices. Organic fish farming is perceived as a natural production method that combines eco-friendliness with fish welfare: '[…] *organic, the fish is happy* […]' (Feucht and Zander 2015). The following summary defines the attributes that consumers associate with organic aquaculture:

- Exclusive breeding of native fish species (Feucht and Zander 2015)
- Pesticide-free (O'Dierno and Myers 2006)
- Medication-/antibiotics-free (O'Dierno and Myers 2006; Feucht and Zander 2015)
- Environmentally friendly (Aarset et al. 2001; O'Dierno and Myers 2006)
- Better taste (O'Dierno and Myers 2006)
- Better animal welfare (Aarset et al. 2001; O'Dierno and Myers 2006)
- Safer (O'Dierno and Myers 2006)
- More nutritious (O'Dierno and Myers 2006)

Organic aquaculture is meant to be a more traditional aquaculture (the term 'fish farming' might apply better to this kind of production), with a low level of technical input. According to these expectations, organic aquaculture should use earth ponds or flow-through systems, and not closed recirculation systems, as these were perceived as too technical and artificial. In the case of sea species, open-sea cages are preferred (Stefani et al. 2011). Sometimes, organic fish farms are assumed to be small to medium sized, as larger production is often associated with industrial livestock farming, which contradicts the idea of organic production (Feucht and Zander 2015). These product attributes and farm characteristics might explain why certified organic seafood has been reported to be preferred by consumers who are particularly concerned about sustainability in their food choices (Zander et al. 2018). In agreement with the findings of Aarset et al. (2004) and Zanoli and Naspetti (2002), among the motives for buying organic fish, there is the avoidance of potential negative consequences associated with the production and consumption of conventional products. However, in most studies, compared to the attributes such as origin, and even relatively unspecific (sustainability) claims, organic certification turns out to be less important (O'Dierno and Myers 2006; Stefani et al. 2011; Mauracher et al. 2013; Zander and Feucht 2017; Zander et al. 2018).

Consumer Preferences and Willingness to Pay for Organic Seafood in Europe

Preferences and willingness to pay are of crucial relevance when defining market opportunities for organic food and seafood. A number of studies have analysed consumer preferences and their willingness to pay (WTP) for seafood with different sustainability attributes (Budak et al. 2006; Olesen et al. 2010; Stefani et al. 2011; Mauracher et al. 2013; Ankamah-Yeboah et al. 2016; Risius et al. 2017; Zander and Feucht 2017; Zander et al. 2018). Ankamah-Yeboah et al. (2016) showed that in Denmark, consumers are willing to pay a price premium of almost 20% for organic salmon compared to conventional salmon. They used real market data from a household panel and carried out a hedonic price analysis. They reported that consumer WTP was higher for seafood with the organic logo than for seafood labelled with the MSC logo. Olesen et al. (2010) reported additional WTP of 15% on average for organic salmon as long as the fish colour was comparable. By using experiments based on consumer choice, they analysed this WTP. According to Risius et al. (2017), smoked trout labelled with the organic 'Naturland' logo was preferred over products with the ASC label.

In a recent contingent valuation study, more than 4000 consumers in 8 European countries were asked about their WTP for different sustainability attributes of farmed fish (Zander and Feucht 2017). The consumers were asked for their additional WTP for seven different product attributes, all of which were related to sustainability: 'sustainably produced', 'organically produced', 'locally produced', 'produced according to higher animal welfare standards' and 'produced in Europe'. On average across all countries, the additional WTP was highest for 'organic production' (+14.8%), followed by 'sustainably produced' (+14%), 'produced with higher animal welfare standards' (+14%), 'locally produced' (+12.6%) and 'produced in Europe' (+9.4%). Thus, organic and sustainable production and also higher animal welfare standards appear to be the most promising attributes with respect to product differentiation in the European fish market.

Consideration of these results by country, which were particularly variable, is more interesting than the overall average (Fig. 3.12). The highest overall level of additional WTP was in Germany, followed by Italy. In Finland, Germany, Spain and the UK, WTP was highest for higher animal welfare standards, while 'organic production' was the most important attribute in France, Ireland and Poland. 'Local origin' was particularly important in Finland, while in all other countries, 'local

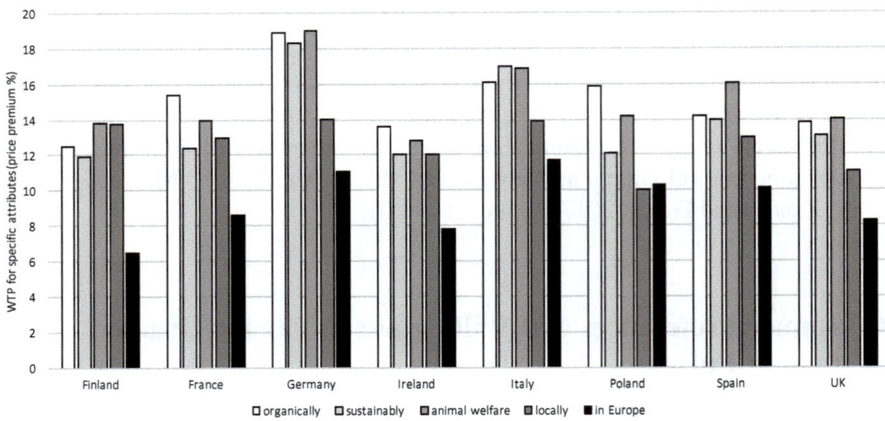

Fig. 3.12 Average additional willingness to pay (in%) for organic fish production compared to the other sustainability attributes indicated. (Source: Adapted from Zander and Feucht (2017))

origin' was outperformed by specific production practices, such as 'sustainably', 'organic' or 'animal welfare' aspects.

These data confirm earlier studies that stressed the importance of animal welfare considerations for consumer demand (O'Dierno and Myers 2006; Kupsala et al. 2013; Feucht and Zander 2015). The preference for local or domestic aquaculture products, which was also reported for previous studies (Stefani et al. 2011; Claret et al. 2012; Mauracher et al. 2013; McClenachan et al. 2016; Risius et al. 2017), is confirmed only for Finland and France. The low additional WTP for 'European origin' confirms earlier data from Pieniak et al. (2013) on consumers in the Czech Republic, Germany, Greece, Italy, Portugal, Romania, Sweden and the UK, although this contradicts Altintzoglou et al. (2010), who reported that indications of European origins enhanced the image of fish.

Figure 3.13 shows the distribution of the consumer WTP in the eight study countries indicated. A very small share of the participants have an additional WTP as high as 100%, as those who would be willing to pay double the price for organic seafood. Some differences become obvious when looking at the share of participants who are willing to pay 50% more: this share is about 10% in Finland but nearer 20% in Germany. About 25% of the Finnish participants and 30% of the German participants were willing to pay a price premium of 20%.

Concluding Remarks

The context of organic aquaculture is very diverse worldwide. Relevant macrotrends indicate:

- Aquaculture production is nearly reaching yields of wild-caught fish.
- A strong positive trend in aquaculture production over the last 10 years.

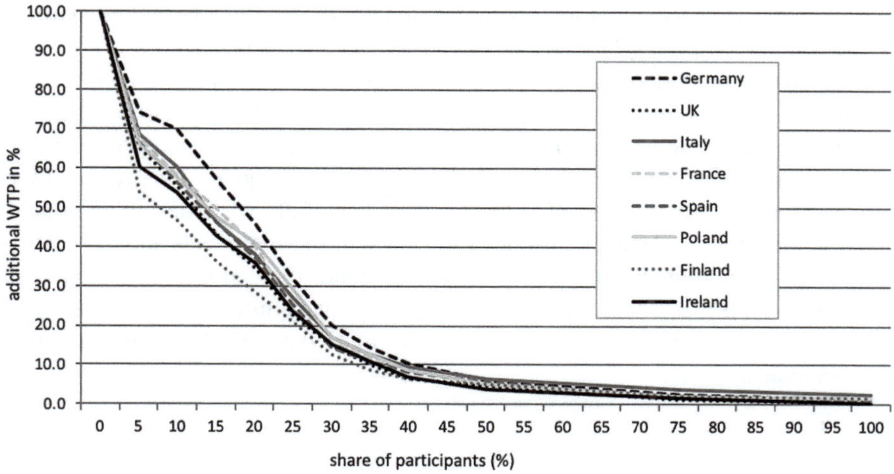

Fig. 3.13 Distribution of consumer willingness to pay (WTP) for 'organically produced' seafood according to country. (Source: own calculations)

- Consumption of organic food is expanding rapidly, particularly in Europe, and organically farmed fish are more in demanded, again, particularly in Europe.
- Organic aquaculture appears to be particularly suitable for developing countries under specific climate conditions, which provides positive stimulus in terms of rural development.

This framework might suggest positive expectations for the expansion of organic aquaculture. However, substantial limitations must be taken into consideration. The data available on the economics of organic farming in Europe show an ambiguous situation, where real profitability relative to conventional aquaculture is dependent on the fish species and the country. Profits are not always guaranteed, and conversion to organic might be an opportunity only for already established farms. The regulatory framework for organic aquaculture appears to be particularly complex, with over 80 national and private standards, and producers from developing countries might face difficulties in marketing their products in western countries. From an economic perspective, the very limited size of the sector implies obstacles for operators, such as diseconomies of scale, limited availability and high prices for purchased inputs, major limitations for processing and high costs of distribution.

The growing demand for organically farmed fish is of course the main solution in the medium term. However, given the potential consumer confusion between wild-caught and organic fish, and between organic labels and other eco-labels, increasing consumer awareness and knowledge through improved marketing and communication is paramount to sustain the growth of this demand. Also, increased supply should result in a decreased price premium (Ankamah-Yeboah et al. 2016). However, as long as the demand for organically farmed fish continues to grow, the price premium is likely to remain relatively high. The limited market size is often associated with low product differentiation and limited product range, which might represent major obstacles for consumers (Castellini et al. 2012).

According to various studies, consumers associate organic aquaculture with small-scale, natural production methods, as preferably in natural ponds and lakes or open-sea cages. Studies show that most consumers do not consider closed recirculation systems to be ‘organic’. Although this is sometimes far from the reality, their concerns and expectations need to be considered carefully when designing organic production systems. As the consumer knowledge of these products and their labelling is low, communication is again very important, to avoid consumer cognitive dissonance. Also, it is important that the organic aquaculture sector maintains and improves its high production standards and aligns these further with consumer expectations.

References

Aarset B, Beckmann S, Bigne J, Beveridge M, Bjorndal T, Bunting MJ, McDonagh P, Mariojouls C, Muir JF, Prothero A, Reisch LA, Smith AP, Tveteras R, Young JA (2001) The European consumers’ understanding and perceptions of organic salmon production. International Institute of Fisheries Economics and Trade

Aarset B, Beckmann S, Bigne E, Beveridge M, Bjorndal T, Bunting J, McDonagh P, Mariojouls C, Muir J, Prothero A, Reisch L, Smith A, Tveteras R, Young J (2004) The European consumers’ understanding and perceptions of the “organic” food regime. Br Food J 106(2):93–105. https://doi.org/10.1108/00070700410516784

Ahmed N, Thompson S, Glaser M (2018) Transforming organic prawn farming in Bangladesh: potentials and challenges. J Clean Prod. Elsevier Ltd 172:3806–3816. https://doi.org/10.1016/j.jclepro.2017.06.110

Almeida C, Altintzoglou T, Cabral H, Vaz S (2015) Does seafood knowledge relate to more sustainable consumption? Br Food J. Emerald Group Publishing Limited 117(2):894–914. https://doi.org/10.1108/BFJ-04-2014-0156

Altintzoglou T, Verbeke W, Vanhonacker F, Luten J (2010) The image of fish from aquaculture among Europeans: impact of exposure to balanced information. J Aquat Food Product Technol. Taylor & Francis Group 19(2):103–119. https://doi.org/10.1080/10498850.2010.492093

Altintzoglou T, Vanhonacker F, Verbeke W, Luten J (2011) Association of health involvement and attitudes towards eating fish on farmed and wild fish consumption in Belgium, Norway and Spain. Aquac Int. Springer Netherlands 19(3):475–488. https://doi.org/10.1007/s10499-010-9363-2

Ankamah-Yeboah I, Nielsen M, Nielsen R (2016) Price premium of organic salmon in Danish retail sale. Ecol Econ. Elsevier B.V. 122:54–60. https://doi.org/10.1016/j.ecolecon.2015.11.028

Ankamah-Yeboah I, Nielsen M, Nielsen R (2017) Market conditions for organic aquaculture: market and price analysis. IFRO Repor. Edited by D. of F. and R. Economics and U. of Copenhagen. Frederiksberg. Available at: http://ifro.ku.dk/publikationer/ifro_serier/rapporter/

Arvanitoyannis IS, Krystallis A, Panagiotaki P, Theodorou AJ (2004) A marketing survey on Greek consumers’ attitudes towards fish. Aquac Int. Kluwer Academic Publishers 12(3):259–279. https://doi.org/10.1023/B:AQUI.0000036137.29397.12

Asche F, Larsen T, Smith MD, Sogn-Grundvag S, Young JA (2013) Pricing of eco-labels for salmon in UK supermarkets. Working paper EE 13-02, Duke environmental and energy economics working paper series

Bellon S, Lamine C (2009) Conversion to organic farming: a multidimensional research object at the crossroads of agricultural and social sciences – a review. Sustain Agric. https://doi.org/10.1007/978-90-481-2666-8_40

Bergleiter S, Berner N, Censkowsky U, Julià-Camprodon G (2009) Organic aquaculture 2009: production and markets. Gräfelfng

Budak F, Budak DB, Kacira OO, Yavuz MC (2006) Consumer willingness to pay for organic sea bass in Turkey. Israeli J Aquac-Bamidgeh 58(2):116–123

Castellini A, De Boni A, Gaviglio A, Mauracher C, Ragazzoni A, Roma R (2012) Prospects and challenges for development of organic fish farming in Italy. New Medit 11(4 SPECIAL):23–26. Available at: https://www.scopus.com/inward/record.uri?eid=2-s2.0-84879752250&partnerID=40&md5=bc498710f10312af3adf80f1807dcfa4.

Claret A, Guerrero L, Aguirre E, Rincón L, Hernández MD, Martínez I, Benito Peleteiro J, Grau A, Rodríguez-Rodríguez C (2012) Consumer preferences for sea fish using conjoint analysis: exploratory study of the importance of country of origin, obtaining method, storage conditions and purchasing price. Food Qual Prefer. Elsevier 26(2):259–266. https://doi.org/10.1016/J.FOODQUAL.2012.05.006

Di Marco P, Petochi T, Marino G, Priori A, Finoia MG, Tomassetti P, Porrello S, Giorgi G, Lupi P, Bonelli A, Parisi G, Poli BM (2017) Insights into organic farming of European sea bass Dicentrarchus labrax and gilthead sea bream *Sparus aurata* through the assessment of environmental impact, growth performance, fish welfare and product quality. Aquaculture. Elsevier B.V. 471:92–105. https://doi.org/10.1016/j.aquaculture.2017.01.012

Disegna M, Mauracher C, Procidano I, Trevisan G (2009) Characteristics of production and consumption of organic trout in Italy. New Medit 8(3 SUPPL):17–26

EUMOFA (2017) EU organic aquaculture. https://doi.org/10.2771/610600

European Commission Directorate-General for Maritime Affairs and Fisheries (2009) Study on the supply and marketing of fishery

FAO (2014) National aquaculture sector overview. FAO Country notes, China

FAOSTAT (2017) FAO yearbook: fishery and aquaculture statistics. 2015, FAO Annual Yearbook. https://doi.org/10.5860/CHOICE.50-5350

Feucht Y, Zander K (2015) Of earth ponds, flow-through and closed recirculation systems – German consumers' understanding of sustainable aquaculture and its communication. Aquaculture. Elsevier B.V. 438:151–158. https://doi.org/10.1016/j.aquaculture.2015.01.005

Grunert KG, Hieke S, Wills J (2014) Sustainability labels on food products: consumer motivation, understanding and use. Food Policy. Pergamon 44:177–189. https://doi.org/10.1016/J.FOODPOL.2013.12.001

Gutierrez A, Thornton T (2014) Can consumers understand sustainability through seafood eco-labels? A U.S. and UK case study. Sustainability. Multidisciplinary Digital Publishing Institute 6(11):8195–8217. https://doi.org/10.3390/su6118195

Home R, Indermuehle A, Tschanz A, Ries E, Stolze M (2018) Factors in the decision by Swiss farmers to convert to organic farming. Renew Agric Food Syst:1–11. https://doi.org/10.1017/S1742170518000121

Honkanen P, Ottar Olsen S (2009) Environmental and animal welfare issues in food choice. Brit Food J. Emerald Group Publishing Limited 111(3):293–309. https://doi.org/10.1108/00070700910941480

IFOAM (2006) Basic standards for organic production and processing (IFOAM). Technol Anal Strateg Manag. ISBN:3-934055-58-3

Kalshoven K, Meijboom FLB (2013) Sustainability at the crossroads of fish consumption and production ethical dilemmas of fish buyers at retail organizations in The Netherlands. J Agric Environ Ethics. Springer Netherlands 26(1):101–117. https://doi.org/10.1007/s10806-011-9367-6

Kupsala S, Jokinen P, Vinnari M (2013) Who cares about farmed fish? Citizen perceptions of the welfare and the mental abilities of fish. J Agric Environ Ethics. Springer Netherlands 26(1):119–135. https://doi.org/10.1007/s10806-011-9369-4

Langer A, Eisend M, Kuss A (2007) The impact of eco-labels on consumers: less information, more confusion? ACR Eur Adv, E-08

Lembo G, Jokumsen A, Spedicato MT, Facchini MT, Bitetto I (2018) Assessing stakeholder's experience and sensitivity on key issues for the economic growth of organic aquaculture production. Mar Policy. Elsevier Ltd 87(October 2017):84–93. https://doi.org/10.1016/j.marpol.2017.10.005

Lernoud J, Willer H (2017) Organic agriculture worldwide: current statistics. In: Willer H, Lernoud J (eds) The world of organic agriculture: statistics and emerging trends 2017. Research Institute of Organic Agriculture (FiBL)/IFOAM-Organics International, Frick/Bonn

Lernoud J, Willer H (2018) Organic agriculture worldwide: current statistics, the world of organic agriculture: statistics and emerging trends 2017. In Willer H, Lernoud J (eds). Research Institute of Organic Agriculture (FiBL), Frick, and IFOAM-Organics International, Bonn. Available at: http://www.organic-world.net/fileadmin/documents/yearbook/2010/world-of-organic-agriculture-2010.pdf

Mauracher C, Tempesta T, Vecchiato D (2013) Consumer preferences regarding the introduction of new organic products. The case of the Mediterranean sea bass (*Dicentrarchus labrax*) in Italy. Appetite 63:84–91. https://doi.org/10.1016/j.appet.2012.12.009

McClenachan L, Dissanayake STM, Chen X (2016) Fair trade fish: consumer support for broader seafood sustainability. Fish Fish. Wiley/Blackwell (10.1111) 17(3):825–838. https://doi.org/10.1111/faf.12148

Mente E, Karalazos V, Karapanagiotidis IT, Pita C (2011) Nutrition in organic aquaculture: an inquiry and a discourse. Aquac Nutr 17(4):e798–e817. https://doi.org/10.1111/j.1365-2095.2010.00846.x

Nair CM, Salin KR, Joseph J, Aneesh B, Geethalakshmi V, New MB (2014) Organic rice-prawn farming yields 20% higher revenues. Agron Sustain Dev. Springer-Verlag France 34(3):569–581. https://doi.org/10.1007/s13593-013-0188-z

Naspetti S, Lampkin N, Nicolas P, Stolze M, Zanoli R (2011) Organic supply chain collaboration: a case study in eight EU countries. J Food Prod Mark 17(2–3):141–162. https://doi.org/10.1080/10454446.2011.548733

Naspetti S, Mandolesi S, Buysse J, Latvala T, Nicholas P, Padel S, Van Loo EJ, Zanoli R (2017) Determinants of the acceptance of sustainable production strategies among dairy farmers: development and testing of a modified technology acceptance model. Sustainability (Switzerland) 9(10). https://doi.org/10.3390/su9101805

Naturland (2017) Naturland standards for organic aquaculture version 5/2017. Gräfelfing. Available at: www.naturland.de

O'Dierno LJ Myers JJ (2006) Consumer perceptions and preferences for organic aquatic products: results from the telephone survey. New Jersey Agricultural Experiment Station, (June)

Olesen I, Alfnes F, Røra MB, Kolstad K (2010) Eliciting consumers' willingness to pay for organic and welfare-labelled salmon in a non-hypothetical choice experiment. Livest Sci. Elsevier B.V. 127(2–3):218–226. https://doi.org/10.1016/j.livsci.2009.10.001

Omoto R, Scott S (2016) Multifunctionality and agrarian transition in alternative agro-food production in the global south: the case of organic shrimp certification in the Mekong Delta, Vietnam. Asia Pac Viewpoint 57(1):121–137. https://doi.org/10.1111/apv.12113

Padel S (2001) Conversion to organic farming: a typical example of the diffusion of an innovation. Sociol Rural 41:40–61

Paul BG, Vogl CR (2012) Key performance characteristics of organic shrimp aquaculture in Southwest Bangladesh. Sustainability 4(5):995–1012. https://doi.org/10.3390/su4050995

Pieniak Z, Vanhonacker F, Verbeke W (2013) Consumer knowledge and use of information about fish and aquaculture. Food Policy. Pergamon 40:25–30. https://doi.org/10.1016/J.FOODPOL.2013.01.005

Prein M, Bergleiter S, Ballauf M, Brister D, Halwar M, Hongrat K, Kahle J, Lasner T, Lem A, Lev O, Morrison C, Shehadeh Z, Stamer A, Wainberg A (2012) Organic aquaculture: the future of expanding niche markets. In: Subasinghe RP, Arthur JR, Bartley DM, De Silva SS, Halwart M, Hishamunda N, Mohan CV, Sorgeloos P (eds) Farming the waters for people and food. FAO/NACA, Rome, pp 549–568

Prins H, Stokkers R, Immink V, Hoste V (2015) Farm economics and competitiveness of organic aquaculture deliverable D3.2. Available at: https://www.oraqua.eu

Risius A, Janssen M, Hamm U (2017) Consumer preferences for sustainable aquaculture products: evidence from in-depth interviews, think aloud protocols and choice experiments. Appetite. Academic Press 113:246–254. https://doi.org/10.1016/J.APPET.2017.02.021

Schlag A, Ystgaard K (2013) Europeans and aquaculture: perceived differences between wild and farmed fish. Br Food J. Emerald Group Publishing Limited 115(2):209–222. https://doi.org/10.1108/00070701311302195

Stefani G, Scarpa R, Cavicchi A (2011) Exploring consumer's preferences for farmed sea bream, (March). https://doi.org/10.1007/s10499-011-9495-z

Stofferahn CW (2009) Personal, farm and value orientations in conversion to organic farming. J Sustain Agric 33(8):862–884. https://doi.org/10.1080/10440040903303595

Stubbe Solgaard H, Yang Y (2011) 'Consumers' perception of farmed fish and willingness to pay for fish welfare. Br Food J. Emerald Group Publishing Limited 113(8):997–1010. https://doi.org/10.1108/00070701111153751

Vanhonacker F, Altintzoglou T, Luten J, Verbeke W (2011) Does fish origin matter to European consumers? Br Food J. Emerald Group Publishing Limited 113(4):535–549. https://doi.org/10.1108/00070701111124005

Verbeke W, Sioen I, Brunsø K, De Henauw S, Van Camp J (2007) Consumer perception versus scientific evidence of farmed and wild fish: exploratory insights from Belgium. Aquac Int. Kluwer Academic Publishers 15(2):121–136. https://doi.org/10.1007/s10499-007-9072-7

Xie B, Qin J, Yang H, Wang X, Wang YH, Li TY (2013) Organic aquaculture in China: a review from a global perspective. Aquaculture. Elsevier B.V. 414–415:243–253. https://doi.org/10.1016/j.aquaculture.2013.08.019

Zander K, Feucht Y (2017) 'Consumers' willingness to pay for sustainable seafood made in Europe. J Int Food Agribusiness Mark:1–25. https://doi.org/10.1080/08974438.2017.1413611

Zander K, Padel S, Zanoli R (2015) EU organic logo and its perception by consumers. Br Food J. Emerald Group Publishing Ltd 117(5):1506–1526. https://doi.org/10.1108/BFJ-08-2014-0298

Zander K, Risius A, Feucht Y, Janssen M, Hamm U (2018) Sustainable aquaculture products: implications of consumer awareness and of consumer preferences for promising market communication in Germany. J Aquat Food Prod Technol Taylor & Francis 27(1):5–20. https://doi.org/10.1080/10498850.2017.1390028

Zanoli R, Naspetti S (2002) Consumer motivations in the purchase of organic food: a means-end approach. Br Food J 104:643–653. https://doi.org/10.1108/00070700210425930.

Zanoli R, Naspetti S, Janssen M, Hamm U (2015) Mediation and moderation in food-choice models: a study on the effects of consumer trust in logo on choice. NJAS – Wageningen J Life Sci 72:41–48. https://doi.org/10.1016/j.njas.2015.01.001

Chapter 4
Genomics Era on Breeding Aquaculture Stocks

Petros V. Martsikalis, Georgios A. Gkafas, Christos Palaiokostas, and Athanasios Exadactylos

Aquaculture Selective Breeding Background

Introduction

The animal- and plant-related agriculture sector has realized over the years tremendous benefits due to breeding programs. Since presently only ~10% of the total production is derived from genetically improved stock (Gjedrem and Robinson 2014), selective breeding can significantly boost the quality of aquaculture products improving traits of economic and welfare importance.

The definition of a breeding program is quite flexible incorporating different breeding and selection practices. Nevertheless, breeding programs are interconnected with pedigree records especially in the last 50 years. Though technically speaking a selection scheme based only on the phenotype of the candidate could be classified as a breeding program, the inclusion of pedigree information was a critical factor behind the worldwide application of selective breeding. The first family-based breeding program in aquaculture originated from Norway in 1975 and proved to be highly successful, reporting 14% of average genetic gain per generation with an estimated 15-fold benefit to cost ratio (Gjedrem 2010). Since that time, several breeding programs have been successfully deployed in aquaculture targeting several traits like growth, fillet yield, deformities, late maturation, and disease resistance.

P. V. Martsikalis · G. A. Gkafas · A. Exadactylos (✉)
Department of Ichthyology and Aquatic Environment, School of Agricultural Sciences, University of Thessaly, Volos, Greece
e-mail: martsikalis@uth.gr; gkafas@uth.gr; exadact@uth.gr

C. Palaiokostas
Department of Animal Breeding and Genetics, Swedish University of Agricultural Sciences, Uppsala, Sweden
e-mail: christos.palaiokostas@slu.se

G. Lembo, E. Mente (eds.), *Organic Aquaculture*,
https://doi.org/10.1007/978-3-030-05603-2_4

Overall, application of selective breeding in aquaculture is less common compared to plant or livestock breeding; this is partly explained because aquaculture is a more recent activity, at least in the industrial form. Additionally, the high fecundity in many aquaculture species (e.g., in sea bream and sea bass) could have created, at least in the early years of aquaculture booming, the false impression that selective breeding is not a priority. Last but not least, another critical factor behind the delayed application of breeding programs in aquaculture lies on the inherent nature of this activity, which constitutes pedigree recording far more challenging than in terrestrial animals and plants.

However, the technical limitations (pedigree analysis) have nowadays been overcome (Palaiokostas et al. 2018), and it is expected that in the following years selective breeding programs will play a more active role toward increasing the global aquaculture production. Particularly in Europe, nowadays it is expected that over 80% of the aquaculture production originates from selective breeding programs aiming to improve several traits simultaneously (Chavanne et al. 2016).

Impact of Aquaculture Breeding Programs

The estimates of genetic gain in aquaculture species for body weight vary from 2.3% to 42% per generation, with an overall average of approximately 14%. The above-average genetic gain for animal growth implies that production can be doubled in six generations as has been the case in tilapia and salmon (Gjedrem and Rye 2016). Regarding the main European aquaculture species, the realized cumulative genetic gain in growth performance over the number of selected generations varies from +65% for turbot to +900% for trout in terms of harvest weight and from +25% for turbot to +200% for trout regarding thermal growth coefficient.

Furthermore, major genetic gains have been recorded for disease resistance with the most highlighted case being the IPNV resistance in Atlantic salmon. In the above, the identification of a major quantitative trait locus (QTL) greatly assisted the industry toward breeding for infectious pancreatic nervous virus (IPNV)-resistant Atlantic salmon (Houston et al. 2008; Moen et al. 2009). In addition, it has to be stressed that in the vast majority of cases growth exhibit positive genetic correlation with disease resistance (Gjedrem and Rye 2016) with the only exceptions originating from shrimp farming (Gitterle et al. 2005)

Aquaculture Breeding Program Design

Different designs of aquaculture breeding programs have been attempted over the years. The initial predominant design based simply on individual selection has gradually given its place in more elaborate schemes in which pedigree information is utilized as well. Additionally, aquaculture breeding programs could also be classified as to whether multiplier stations are utilized in order to enhance commercial production. However, the practice of multiplying stations is characteristic of salmonid breeding programs, while in contrast it is not common in Mediterranean aquaculture. Interestingly, a recent cost-benefit analysis for aquaculture breeding program designs based on gilthead sea bream identified as the most profitable the design without a multiplier station (Janssen et al. 2018). However, we also need to stress the biological differences regarding reproduction and fecundity levels between the salmonids and the main Mediterranean aquaculture marine fish like the gilthead sea bream and the European sea bass.

Inbreeding Management

The risks of negative performance elevate significantly when inbreeding rate is above 1% per generation (Meuwissen and Wooliams 1994). Maintaining a breeding nucleus of high genetic diversity is therefore essential for every breeding program. Particularly for organic aquaculture, it would be essential to strive for greater numbers of contributing sires and dams in each generation of the breeding program. As a rule of thumb, a minimum size of at least 50 broodfish should be guaranteed to minimize potential negative effects due to inbreeding (Gjedrem and Robinson 2014). However, this last number represents an absolute minimum, with numbers above 100 being advisable for guaranteeing the long-term sustainability of the breeding program. This is further documented from data coming from aquaculture breeding programs in Europe which showed that the number of broodfish involved in each generation greatly varies from less than 100 to more than 800 (Chavanne et al. 2016).

Estimation of Breeding Values

Developments in statistical tools were crucial for the successful implementation of breeding programs. A most crucial point was the framework establishment of statistical models which allowed the efficient estimation of breeding values like the best linear unbiased prediction (BLUP) methodology (Henderson 1975). In its traditional form, BLUP utilizes pedigree information circumventing the limitation of classic linear models regarding the maximum number of estimated parameters. Additionally, BLUP allows breeding value estimation by utilizing the pedigree information even for animals with no phenotypic records. Obviously, reliable pedigree recordings are of paramount importance for the accurate calculation of breeding values.

Nevertheless, limitations of BLUP are apparent especially for traits where no phenotypic records are available for the selection candidates per se (e.g., disease challenge experiment where records are only available for the sibs of the candidate) since BLUP utilizes only the between-family genetic variation. The modern approach upon which accuracy improvement of the estimated breeding values is based utilizes information from genetic markers allowing for higher resolution of the actual relationship between selection candidates. In recent years, both animal and plant breeding witnessed an immense amount of research aiming at increasing the accuracy of breeding values using genome-wide genetic markers. Termed as genomic selection (Meuwissen et al. 2001), this has been widely adopted both by industry and academia. Particularly, for aquaculture the usage of genetic markers also allows the circumvention of the prior necessity of separate tank rearing for each family of the breeding program.

Important Selection Traits for Organic Aquaculture

Feed Efficiency

Feed costs normally constitute the highest running expenses in aquaculture. Excluding shellfish farming, feed costs in an intensive aquaculture activity would typically account for approximately 50–60% of the total running costs. Therefore, selection for improved feed efficiency would greatly benefit the industry, but this has proven to be a challenging task mainly due to the difficulty of obtaining accurate phenotypic measurements that are actually representative of the trait (Daulé et al. 2014). Prior experience demonstrated that selection for higher growth could favor animals with high feed intake. However, the last does not necessarily guarantee that these animals have the highest genetic value for feed efficiency. Since feed efficiency positively correlates with feed intake, up to a threshold, a suboptimal feeding strategy would easily result in increased feed intake and reduced feed efficiency (de Verdal et al. 2017). Particularly in organic aquaculture, selecting for increased feed efficiency is of paramount importance, since apart from the obvious economic benefits, an increase of feed efficiency would simultaneously result in a reduced environmental impact through the proportional reduction of wasted (not consumed) feed.

Disease Resistance

Disease outbreaks pose serious challenges for any animal- or plant-related activity and particularly in aquaculture where in many occasions efficient therapeutic agents are lacking. This is even more relevant in the case of organic farming where there are restrictions in the use of antibiotics and the available therapeutic agents are even more limited. In addition, the inherent environment of an aquaculture activity (e.g., net pens) is a great challenge to tackle disease outbreaks. Here again, selective breeding might be a valuable tool to prevent the detrimental effects of disease outbreaks (Bishop and Woolliams 2014). For example, in shrimp farming where the species innate immune system constitutes vaccines inefficient, selective breeding together with strict biosecurity and effective management practices are the only tools available for handling disease outbreaks.

On the positive side, moderate to high heritabilities for disease resistance have been reported in numerous aquaculture species, indicating that rapid genetic progress could be achieved through selective breeding (Ødegård et al. 2011). Particularly for disease resistance traits, approaches utilizing genomic information have proven invaluable. Several studies have demonstrated the potential of breeding for disease resistance like for IPNV (Houston et al. 2008; Moen et al. 2009), pancreas disease (Gonen et al. 2015), and sea lice (Ødegård et al. 2014; Tsai et al. 2016; Correa et al. 2017) in Atlantic salmon. Additional studies have been also conducted in rainbow trout (Vallejo et al. 2016, 2017), gilthead sea bream (Palaiokostas et al. 2016), catfish (Zhou et al. 2017), Asian sea bass (Wang et al. 2017), and European sea bass (Palaiokostas et al. 2018).

Selection for Skewed Sex Ratio Stocks

Phenotypic sex is generally labile in fish with a number of aquaculture fish species exhibiting sexual dimorphism in a range of traits of interest such as age at maturity. Some of the most important cases with considerable sexual dimorphism regarding growth include the Nile tilapia and European sea bass. In the case of Nile tilapia, sexual dimorphism favors males (Mair et al. 1997), and several studies have identified genomic regions with strong influence in sex determination (Lee et al. 2011; Eshel et al. 2012; Palaiokostas et al. 2013, 2015; Wessels et al. 2017), demonstrating that despite the complexity of the trait, selective breeding could be of value. Additionally, taking into account that hormonal sex reversal is still widely practiced in tilapia farming, selective breeding appears as a necessity. In the case of European sea bass, females demonstrate superior growth, and sex determination is suspected to be polygenic with a particularly high heritability of 0.62 (Vandeputte et al. 2007), implying that there is plenty of margin upon which selection can achieve a rapid and substantial improvement.

Genotype by Environment Interaction (GxE) in Selective Breeding Programs

The GxE interactions simply imply that organic, as well as conventional, fish derived from the single nucleus of a selective breeding program might show lower-than-expected genetic gains in different environmental conditions, e.g., other organic/conventional aquaculture sites. Therefore, it is imperative to understand the key contribution of potential GxE interactions among different environments and their implications to the economic and sustainable development of commercial aquaculture.

GxE interaction is defined as the difference in the response of genotypes across various macro-environments (Hammond 1947; Falconer and Mackay 1996; Lynch and Walsh 1998; Lin and Togashi 2002). When the existence of GxE interaction is confirmed, we are expecting an unpredictable variation of a single genotype performance among different environments (Evans and Langdon 2006). GxE interaction can appear in two forms: first, as heterogeneity of genetic variance, where there is a difference in genetic variance concerning a trait across various environments (Falconer and Mackay 1996; Lynch and Walsh 1998; Ponzoni et al. 2008), and second as genotype re-ranking across different environments. The latter is the most serious since it implies that selection gains could significantly deviate from the expected ones.

Moderate re-ranking has been observed for survival traits (mean genetic correlation 0.54), while re-ranking was lower for traits like age at maturity and appearance (mean genetic correlation 0.86). However, most of the recorded genetic correlations on different environments are positive, implying that performance on a different environment would still infer an advantage though potentially of different magnitude from the expected one (Sae-Lim et al. 2016). The presence of GxE interactions in selective breeding programs was detected in several aquaculture species such as rainbow trout (Kause et al. 2003; Sae-Lim et al. 2014), Atlantic cod (Kolstad et al. 2006), Arctic charr (Nilsson et al. 2010; Chiasson et al. 2013), shrimp (Li et al. 2015; Lu et al. 2017), Asian and European sea bass (Saillant et al. 2006; Dupont-Nivet et al. 2010; Domingos et al. 2013; Le Boucher et al. 2013), gilthead sea bream (Navarro et al. 2009), turbot (Guan et al. 2016), sole (Mas-Muñoz et al. 2013), and Nile tilapia (Bentsen et al. 2012; Trong et al. 2013; Turra et al. 2016).

Taking into account the obtained results of the aforementioned studies, it can be concluded that the GxE interaction effect seems to be more significant (in terms of commercial aquaculture) when the production environments present substantial differences in environmental factors such as their hydrological and physicochemical attributes. This fact was also supported by Luan et al. (2008), who confirmed the existence of GxE interaction in traits between reared fish populations in freshwater and brackish water, respectively. In contrast, the GxE interaction was nonexistent when there was a great extent of similarity between the rearing environments (Nguyen 2016; Sae-Lim et al. 2016).

Consequently, the assessment of the GxE interaction magnitude through the quantification of genotype re-ranking effect is critical for the proper design and optimization of a selective breeding program in aquaculture. Therefore a breeding program should establish separate test stations across different production environments simulating the specific environmental conditions of the target markets or in the case of strong re-ranking occurrence establish separate breeding programs focusing on the site-specific environmental conditions of each production environment. Nevertheless, all of the above could only be possible if justified by a cost-benefit study. Taking into account the usually considerable running costs of a breeding program, it could be particularly hard to justify the need for simultaneously running additional ones. A potential solution would be the creation and funding of European/national breeding programs in collaboration with the private sector.

Novel Omics Technologies in Organic Aquaculture

Recent genomic experiments have established a new potential in aquaculture industry. New methodologies and advanced state-of-the-art facilities have provided a better understanding of the farmed species molecular profile. Next-generation sequencing (NGS) techniques have resulted in the annotation of many teleost genomes of interest, and omics are becoming a powerful multidisciplinary strategy in aquaculture programs. These methodologies include *genomics* to study DNA variations, *transcriptomics* to study gene expression, *proteomics* to assess protein expression, and *metabolomics* for metabolite variability.

The produced huge amount of data by omics techniques may require weeks or even months to analyze and process data generated. The demand for high-performance computing in terms of speed and reliability is subject to the increasing data in size through the NGS techniques of the last decade. High-performance clusters have emerged lately as promising technology drivers in the "big data" era. Due to the complexity of these biological big data, bioinformatics skills are required, bringing together multidisciplinary fields such as computer science, biology, mathematics, medicine, etc. Nowadays, genomic studies benefit of high-performance computing (HPC) in conjunction with parallelism infrastructures, providing bioinformatically speaking interesting opportunities for research initiatives.

Genomics

A number of various genomic and cost-efficient tools have been developed in the last decade. Restriction site-associated DNA sequencing (RAD-seq) and shotgun sequencing are widely used in genomic-wide association studies (GWAS) providing relatively low-cost discovery and genotyping of thousands of single nucleotide polymorphisms (SNPs) (Robledo et al. 2017). SNPs are becoming the marker of choice in genetic selection programs, due to their direct association between fitness-related traits and genotype. In farmed stocks, variability provides a key role tool to study genetic diversity with regard to differential fitness-related traits. The latter relationship is called heterozygosity-fitness correlation. The definition of a heterozygosity-fitness correlation (after David 1998) is "the empirical observation of a correlation between heterozygosity measured at a marker locus, or at a set of marker loci, and a fitness-related trait." Three main hypotheses have been proposed to explain the heterozygosity-fitness correlation: (a) the direct effect hypothesis, i.e., heterozygote advantage due to overdominance at the specific locus scored; (b) the local effect hypothesis, i.e., heterozygote advantage detected at marker loci that are closely linked to fitness loci; and (c) the general effect hypothesis, i.e., heterozygote advantage due to genome-wide effects (high level of heterozygosity at the marker loci reflecting a high level of heterozygosity in the genome as a whole).

The apparent genotype-fitness association mapping is becoming crucial for additional genome scans for quantitative trait loci (QTL) in marker-assisted selection (MAS) programs. However, more recent studies into other traits in aquaculture have highlighted that polygenic architecture is indeed the norm in aquaculture species too, with each QTL explaining only a very small percentage of the genetic variation, further indicating that MAS is likely to have very limited success (Palaiokostas and Houston 2018).

Transcriptomics

Transcriptomics comprises one of the crucial assessments for the improvement of husbandry practices in aquaculture (Chapman et al. 2014). Nutrition is one of the most important phases in an aquaculture scheme, and therefore feed efficiency should be optimized. This is often subject to the expression of genes involved, and the analysis of transcripts is a fundamental methodology to quantify the expression level change under different conditions (Kavouras et al. 2017). Various methodologies have been developed for this purpose based on hybridization (e.g., high-density oligo-microarrays) or sequencing (e.g., Sanger sequencing of complementary DNA (cDNA) or expressed sequence tag (EST)) using real-time PCR. The main limitations of these techniques are either the reliance upon existing knowledge for genome sequence or expensive and generally not quantitative (see review in Wang et al. 2009). Nowadays, RNA-sequence (RNA-seq) allows exploring genomic resources profiles using NGS to reveal a far more precise measurement of levels of presence and quantity of transcripts to a better understanding of the development of an alteration in phenotype.

Proteomics

High-throughput proteomics can be a useful tool to determine protein expression profiles in exploring unresolved questions such as host-pathogen interactions at the individual level with respect to fitness association analyses. More broadly, it is important to understand the consequences of host-pathogen interactions in the context of other regulatory mechanisms, such as nutrition, reproduction, competition, and behavioral factors, as well as ecological/environmental factors, such as habitat use and climate changes (Irvine 2006). In general, fitness impacts due to pathogen load can also impact on demographics and therefore indirectly on factors that can determine genetic diversity, leading to the identification of new vaccine targets and novel therapeutics in marine aquaculture.

Metabolomics

Innovative methodologies, such as *metabolomics*, are emerging to discover physiological responses to their altering biotic and abiotic environments. Their applications hold a great potential to gain insight into mechanisms with valuable information to recognize the metabolites present within tissues and organs. Metabolite profiling, for example, in three different gilthead sea bream (*Sparus aurata*) aquaculture systems shows different levels of glycogen (stress indicator), histidine, alanine, and glycine, using chemometrics and H-1 NMR technique (Savorani et al. 2010). Moreover, a decrease in succinate and an increase in alanine were observed after water exchange in the metabolic profile of hatchery-reared mussel (*Perna canaliculus*) larvae (Young et al. 2016).

Conclusion

Particular attention should be paid to the continuous enhancement and development of the scientific knowledge in the field of selective breeding, so that the EU legislative framework on organic aquaculture is always up to new challenges toward sustainability. Bearing in mind the potential existence of GxE interactions, the genotypes selected in a conventional aquaculture site through a MAS program are often not optimal for organic aquaculture conditions. Therefore, we suggest that MAS programs can be successful provided that the genotypes selection will focus on their performance under site-specific organic farming conditions. This involves the creation and funding of European/national breeding programs in collaboration with the private sector.

Moreover, the selection criteria for traits apart from feed efficiency, product quality, and sexual differentiation should include vitality and resistance to diseases. Potential alternatives to the use of antibiotics could be the adoption of homeopathic treatments. Hence, the aquaculture industry has to make efforts to adopt selective breeding programs which will further promote fish performance and robustness increasing their tolerance against environmental changes. Nevertheless, economic growth in organic aquaculture can be acceptable in condition that fish welfare, environmental protection, and sustainability are not neglected.

The implementation of production rules and best practices for the organic aquaculture should be regularly reviewed, and in case potential issues should arise, these should be addressed by the scientific community in close cooperation with pertinent stakeholders and public authority, in order to ensure the sustainable development of organic aquaculture and consequently increase the organic logo credibility for the consumers.

References

Bentsen HB, Gjerde B, Nguyen NH, Rye M, Ponzoni RW, Palada de Vera MS et al (2012) Genetic improvement of farmed tilapias: genetic parameters for body weight at harvest in Nile tilapia (*Oreochromis niloticus*) during five generations of testing in multiple environments. Aquaculture 338–341:56–65

Bishop SC, Woolliams JA (2014) Genomics and disease resistance studies in livestock. Livest Sci 166:190–198

Chapman RW, Reading BJ, Sullivan CV (2014) Ovary transcriptome profiling via artificial intelligence reveals a transcriptomic fingerprint predicting egg quality in striped bass, *Morone saxatilis*. PLoS One 9:e96818

Chavanne H, Janssen K, Hofherr J, Contini F, Haffray P et al (2016) A comprehensive survey on selective breeding programs and seed market in the European aquaculture fish industry. Aquac Int 24:1287–1307

Chiasson M, Quinton M, Pelletier C, Danzmann R, Ferguson M (2013) Family x environment interactions in the growth and survival of Arctic charr (*Salvelinus alpinus*) grown in brackish and fresh water. Aquac Res 45:1953–1963

Correa K, Bangera R, Figueroa R, Lhorente JP, Yanez JM (2017) The use of genomic information increases the accuracy of breeding value predictions for sea louse (*Caligus rogercresseyi*) resistance in Atlantic salmon (*Salmo salar*). Genet Sel Evol: GSE 49:15

Daulé S, Vandeputte M, Vergnet A, Guinand B, Grima L et al (2014) Effect of selection for fasting tolerance on feed intake, growth and feed efficiency in the European sea bass *Dicentrarchus labrax*. Aquaculture 420–421:S42–S49

David P (1998) Heterozygosity-fitness correlations: new perspectives on old problems. Heredity 80:531–537

de Verdal H, Komen H, Quillet E, Chatain B, Allal F et al (2017) Improving feed efficiency in fish using selective breeding: a review. Rev Aquacult 10:833–851

Domingos JA, Smith-Keune C, Robinson N, Loughnan S, Harrison P, Jerry DR (2013) Heritability of harvest growth traits and genotype-environment interactions in barramundi, *Lates calcarifer* (Bloch). Aquaculture 402–403:66–75

Dupont-Nivet M, Karahan-Nomm B, Vergnet A, Merdy O, Haffray P, Chavanne H et al (2010) Genotype by environment interactions for growth in European seabass (*Dicentrarchus labrax*) are large when growth rate rather than weight is considered. Aquaculture 306:365–368

Eshel O, Shirak A, Weller JI, Hulata G, Ron M (2012) Linkage and physical mapping of sex region on LG23 of Nile Tilapia (*Oreochromis niloticus*). G3 Genes|Genome|Genetics 2:35–42

Evans S, Langdon C (2006) Effects of genotype x environment interactions on the selection of broadly adapted Pacific oysters (*Crassostrea gigas*). Aquaculture 261:522–534

Falconer, D. S., Mackay, T. F. C., 1996 Introduction to quantitative genetics. Longman, Essex

Gitterle T, Salte R, Gjerde B, Cock J, Johansen H et al (2005) Genetic (co)variation in resistance to white spot syndrome virus (WSSV) and harvest weight in *Penaeus (Litopenaeus) vannamei*. Aquaculture 246:139–149

Gjedrem T (2010) The first family-based breeding program in aquaculture. Rev Aquac 2:2–15

Gjedrem T, Robinson N (2014) Advances by selective breeding for aquatic species: a review. Agric Sci 5:1152–1158

Gjedrem T, Rye M (2016) Selection response in fish and shellfish: a review. Rev Aquacult 10:168–179

Gonen S, Baranski M, Thorland I, Norris A, Grove H et al (2015) Mapping and validation of a major QTL affecting resistance to pancreas disease (salmonid alphavirus) in Atlantic salmon (*Salmo salar*). Heredity:1–10

Guan J, Hu Y, Wang M, Wang W, Kong J, Luan S (2016) Estimating genetic parameters and genotype-by-environment interactions in body traits of turbot in two different rearing environments. Aquaculture 450:321–327

Hammond J (1947) Animal breeding in relation to nutrition and environmental conditions. Biol Rev 22:195–213

Henderson CR (1975) Best linear unbiased estimation and prediction under a selection model. Biometrics 31:423–447

Houston RD, Haley CS, Hamilton A, Guy DR, Tinch AE et al (2008) Major quantitative trait loci affect resistance to infectious pancreatic necrosis in Atlantic salmon (*Salmo salar*). Genetics 178:1109–1115

Irvine RJ (2006) Parasites and the dynamics of wild mammal populations. Anim Sci 82:775–781

Janssen K, Saatkamp H, Komen H (2018) Cost-benefit analysis of aquaculture breeding programs. Genet Sel Evol 50:2

Kause A, Ritola O, Paananen T, Mantysaari E, Eskelinen U (2003) Selection against early maturity in large rainbow trout *Oncorhynchus mykiss*: the quantitative genetics of sexual dimorphism and genotype-by-environment interactions. Aquaculture 228:53–68

Kavouras M, Tsilika K, Exadactylos A (2017) A computer algebra system approach in gene expression analysis. Prog Ind Ecol 11(1):49–60

Kolstad K, Thorland I, Refstie T, Gjerde B (2006) Genetic variation and genotype by location interaction in body weight, spinal deformity and sexual maturity in Atlantic cod (*Gadus morhua*) reared at different locations off Norway. Aquaculture 259:66–73

Le Boucher R, Vandeputte M, Dupont-Nivet M, Quillet E, Ruelle F, Vergnet A et al (2013) Genotype by diet interactions in European sea bass (*Dicentrarchus labrax* L.): nutritional challenge with totally plant-based diets. J Anim Sci 91:44–56. https://doi.org/10.2527/jas.2012-5311.

Lee B-Y, Coutanceau J-P, Ozouf-Costaz C, D'Cotta H, Baroiller J-F et al (2011) Genetic and physical mapping of sex-linked AFLP markers in Nile tilapia (*Oreochromis niloticus*). Mar Biotechnol 13:557–562

Li W, Luan S, Luo K, Sui J, Xu X, Tan J, Kong J (2015) Genetic parameters and genotype by environment interaction for cold tolerance, body weight and survival of the Pacific white shrimp *Penaeus vannamei* at different temperatures. Aquaculture 441:8–15. https://doi.org/10.1016/j.aquaculture.2015.02.013

Lin CY, Togashi K (2002) Genetic improvement in the presence of genotype by environment interaction. J Anim Sci 73:3–11

Lu X, Luan S, Cao B, Meng X, Sui J, Dai P et al (2017) Estimation of genetic parameters and genotype-by-environment interactions related to acute ammonia stress in Pacific white shrimp (*Litopenaeus vannamei*) juveniles at two different salinity levels. PLoS One 12(3):e0173835. https://doi.org/10.1371/journal.pone.0173835

Luan TD, Olesen I, Ødegård J, Kolstad K, Dan NC (2008) Genotype by environment interaction for harvest body weight and survival of Nile tilapia (*Oreochromis niloticus*) in brackish and fresh water ponds. In: Proceedings of the 8th international symposium on Tilapia in aquaculture. Cairo, Egypt, October 12–14, 2008. Cited 21 January 2013. Available from URL: http://ag.arizona.edu/azaqua/ista/ISTA8/ProceedingsISTA8.htm

Lynch M, Walsh B (1998) Genetics and analysis of quantitative traits. Sinauer Associates, Sunderland

Mair GC, Abucay JS, a Abella T, a Beardmore J, Skibinski DOF (1997) Genetic manipulation of sex ratio for the large-scale production of all-male tilapia *Oreochromis niloticus*. Can J Fish Aquat Sci 54:396–404

Mas-Muñoz J, Blonk R, Schrama JW, van Arendonk J, Komen H (2013) Genotype by environment interaction for growth of sole (*Solea solea*) reared in an intensive aquaculture system and in a semi-natural environment. Aquaculture 410–411:230–235

Meuwissen THE, Woolliams JA (1994) Effective sizes of livestock populations to prevent a decline in fitness. Theor Appl Genet 89(7–8):1019–1026

Meuwissen THE, Hayes BJ, Goddard ME (2001) Prediction of total genetic value using genome-wide dense marker maps. Genetics 157:1819–1829

Moen T, Baranski M, Sonesson AK, Kjøglum S (2009) Confirmation and fine-mapping of a major QTL for resistance to infectious pancreatic necrosis in Atlantic salmon (*Salmo salar*): population-level associations between markers and trait. BMC Genomics 10:368

Navarro A, Zamorano MJ, Hildebrandt S, Ginés R, Aguilera C, Afonso JM (2009) Estimates of heritabilities and genetic correlations for growth and carcass traits in gilthead seabream (*Sparus auratus* L.), under industrial conditions. Aquaculture 289:225–230

Nguyen NH (2016) Genetic improvement for important farmed aquaculture species with a reference to carp, tilapia and prawns in Asia: achievements, lessons and challenges. Fish Fish 17:483–506. https://doi.org/10.1111/faf.12122

Nilsson J, Brännäs E, Eriksson L-O (2010) The Swedish Arctic charr breeding programme. Hydrobiologia 650:275–282

Ødegård J, Baranski M, Gjerde B, Gjedrem T (2011) Methodology for genetic evaluation of disease resistance in aquaculture species: challenges and future prospects. Aquac Res 42:103–114

Ødegård J, Moen T, Santi N, Korsvoll SA, Kjøglum S et al (2014) Genomic prediction in an admixed population of Atlantic salmon (*Salmo salar*). Front Genet 5:402

Palaiokostas C, Houston R (2018) Genome-wide approaches to understanding and improving complex traits in aquaculture species. CAB Rev: Perspect Agric Vet Sci Nutr Nat Res. https://doi.org/10.1079/pavsnnr201712055

Palaiokostas C, Bekaert M, Khan MGQ, Taggart JB, Gharbi K et al (2013) Mapping and validation of the major sex-determining region in Nile Tilapia (*Oreochromis niloticus* L.) using RAD sequencing. PLoS One 8:e68389

Palaiokostas C, Bekaert M, Khan MGQM, Taggart JBJ, Gharbi K et al (2015) A novel sex-determining QTL in Nile tilapia (*Oreochromis niloticus*). BMC Genomics 16:171

Palaiokostas C, Ferraresso S, Franch R, Houston RD, Bargelloni L (2016) Genomic prediction of resistance to pasteurellosis in gilthead sea bream (*Sparus aurata*) using 2b-RAD sequencing. G3 6(11):3693–3700

Palaiokostas C, Cariou S, Bestin A, Bruant J-S, Haffray P et al (2018) Genome-wide association and genomic prediction of resistance to viral nervous necrosis in European sea bass (*Dicentrarchus labrax*) using RAD sequencing. Genet Sel Evol 50:30

Ponzoni RW, Nguyen NH, Khaw HL, Ninh NH (2008) Accounting for genotype by environment interaction in economic appraisal of genetic improvement programs in common carp *Cyprinus carpio*. Aquaculture 285:47–55

Robledo D, Palaiokostas C, Bargelloni L, Martínez P, Houston R (2017) Applications of genotyping by sequencing in aquaculture breeding and genetics. Rev Aquac 0:1–13

Sae-Lim P, Komen H, Kause A, Mulder H (2014) Identifying environmental variables explaining genotype-by-environment interaction for body weight of rainbow trout (*Oncorhynchus mykiss*): reaction norm and factor analytic models. Genet Sel Evol 46:16

Sae-Lim P, Gjerde B, Nielsen HM, Mulder H, Kause A (2016) A review of genotype-by-environment interaction and micro-environmental sensitivity in aquaculture species. Rev Aquac 8:369–393

Saillant E, Dupont-Nivet M, Haffray P, Chatain B (2006) Estimates of heritability and genotype-environment interactions for body weight in sea bass (*Dicentrarchus labrax* L.) raised under communal rearing conditions. Aquaculture 254:139–147

Savorani F, Picone G, Badiani A, Fagioli P, Capozzi F, Engelsen S (2010) Metabolic profiling and aquaculture differentiation of gilthead sea bream by 1H NMR metabonomics. Food Chem 120:907. https://doi.org/10.1016/j.foodchem.2009.10.071

Trọng TQ, Mulder HA, van Arendonk JAM, Komen H (2013) Heritability and genotype by environment interaction estimates for harvest weight, growth rate, and shape of Nile tilapia (*Oreochromis niloticus*) grown in river cage and VAC in Vietnam. Aquaculture 384–387:119–127

Tsai H-Y, Hamilton A, Tinch AE, Guy DR, Bron JE et al (2016) Genomic prediction of host resistance to sea lice in farmed Atlantic salmon populations. Genet Sel Evol 48:47

Turra EM, Toral FLB, Alvarenga ER, Raidan FSS, Fernandes AFA, Alves GFO (2016) Genotype × environment interaction for growth traits of Nile tilapia in biofloc technology, recirculating water and Cage systems. Aquaculture 460:98–104. https://doi.org/10.1016/j.aquaculture.2016.04.020

Vallejo RL, Leeds TD, Fragomeni BO, Gao G, Hernandez AG et al (2016) Evaluation of genome-enabled selection for bacterial cold water disease resistance using progeny performance data in rainbow trout: insights on genotyping methods and genomic prediction models. Front Genet 7:96

Vallejo RL, Leeds TD, Gao G, Parsons JE, Martin KE et al (2017) Genomic selection models double the accuracy of predicted breeding values for bacterial cold water disease resistance compared to a traditional pedigree-based model in rainbow trout aquaculture. Genet Sel Evol 49:17

Vandeputte M, Dupont-Nivet M, Chavanne H, Chatain B (2007) A polygenic hypothesis for sex determination in the European sea bass *Dicentrarchus labrax*. Genetics 176:1049–1057

Wang Z, Gerstein M, Snyder M (2009) RNA-Seq: a revolutionary tool for transcriptomics. Nat Rev Genet 10(1):57–63

Wang L, Liu P, Huang S, Ye B, Chua E et al (2017) Genome-wide association study identifies loci associated with resistance to viral nervous necrosis disease in Asian seabass. Mar Biotechnol 19:255–265

Wessels S, Krause I, Floren C, Schütz E, Beck J et al (2017) ddRADseq reveals determinants for temperature-dependent sex reversal in Nile tilapia on LG23. BMC Genomics 18:531

Young T, Alfaro AC, Villas-Bôas SG (2016) Metabolic profiling of mussel larvae: effect of handling and culture conditions. Aquac Int 24:843. https://doi.org/10.1007/s10499-015-9945-0

Zhou T, Liu S, Geng X, Jin Y, Jiang C et al (2017) GWAS analysis of QTL for enteric septicemia of catfish and their involved genes suggest evolutionary conservation of a molecular mechanism of disease resistance. Mol Gen Genomics 292:231–242

Chapter 5
Early Life Stages and Weaning

Alicia Estévez, Nikos Papandroulakis, Mathieu Wille, and Patrick Sorgeloos

Development, Feeding, and Nutritional Requirements of Marine and Freshwater Fish Larvae

High differences exist between freshwater and marine fish in terms of reproduction, larval development, nutritional requirements and feeds, as well as in culture methods.

Marine Fish Larval Development, Feeding, and Nutritional Requirements

In the case of marine fish, reproduction involves the collection of large numbers of small (average 1–2 mm diameter) buoyant eggs which, after being released into the water and fertilized, drift to the surface with no parental attention. Newly hatched larvae, also known as eleuthero embryo or prelarvae, carry an elliptical yolk sac containing an oil globule that functions as nutrient store (mostly glycogen and free amino acids in the yolk and triacylglycerol in the oil globule) for further

A. Estévez (✉)
IRTA Sant Carles de la Ràpita, Sant Carles de la Ràpita, Spain
e-mail: alicia.estevez@irta.cat

N. Papandroulakis
Institute Marine Biology, Biotechnology and Aquaculture, Hellenic Center for Marine Research, Heraklion, Crete, Greece
e-mail: npap@hcmr.gr

M. Wille · P. Sorgeloos
Faculty of Bioscience Engineering Laboratory of Aquaculture & Artemia Campus Coupure – Blok F Reference Center, University of Ghent, Ghent, Belgium
e-mail: Mathieu.Wille@UGent.be; Patrick.Sorgeloos@UGent.be

G. Lembo, E. Mente (eds.), *Organic Aquaculture*,
https://doi.org/10.1007/978-3-030-05603-2_5

development of the larva. At hatching the larva is very little developed, it does not have functional eyes, and the digestive system is just a histologically undifferentiated tube (Govoni et al. 1986). After a short period of endogenous nutrition using its yolk reserves, the larva develops the vision and digestive system, the eyes become pigmented, the jaw and mouth are functional, and the digestive system includes a tube-like alimentary canal, liver, pancreas, and gall bladder. Feeding starts on live prey, a process which depends on zooplankton density. As the larvae progress toward becoming a juvenile, the digestive system will continue to differentiate with the formation of pyloric caeca and differentiation of the stomach during the metamorphosis.

Apart from the small size and the undeveloped digestive system, marine fish larvae have a high requirement for several fatty acids, considered essential (EFA), for a proper development and growth. Among these fatty acids, docosahexaenoic acid (DHA, 226n-3) is needed for neural and visual development (Mourente and Tocher 1992); eicosapentaenoic acid (EPA, 20:5n-3), and in general all the omega-3 fatty acids, is needed for growth; and arachidonic acid (ARA, 20:4n-6) is needed for eicosanoid production and plays an important role in the pigmentation processes of flatfish (Estevez et al. 1997; Villalta et al. 2005) and in the resistance to handling stress (Koven et al. 2001). Marine fish larvae cannot elongate or desaturate the precursor fatty acids α-linolenic (ALA, 18:3n-3) and α-linoleic (LA, 18:2n-6) into these long chain fatty acids, and they need to be provided in the diet (Figs. 5.1).

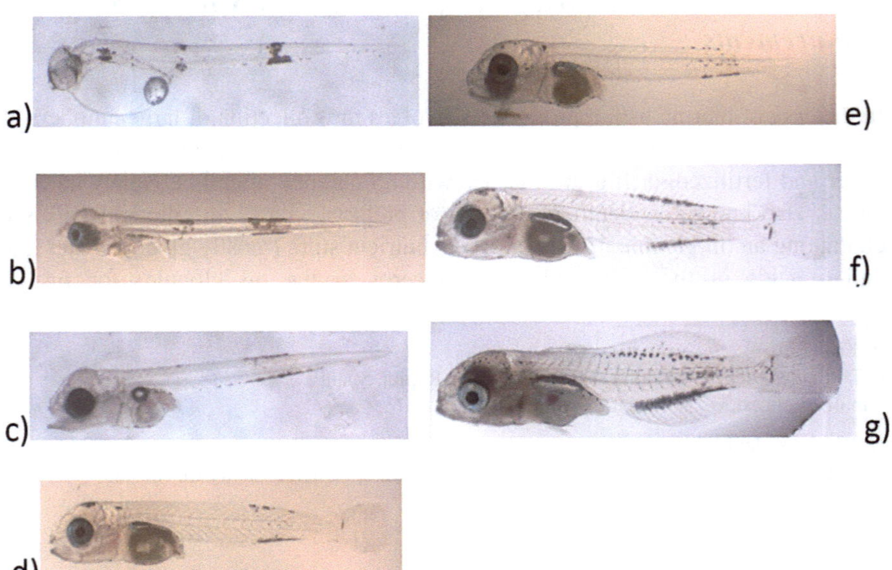

Fig. 5.1 Larval development of gilthead sea bream *Sparus aurata* from hatching (**a**) to 50 days post-hatch (**g**) showing mouth opening (**b**), swim bladder inflation (**c**, 10 dph), flexion (**d**, 20 dph), and tail (**e**) and fins (**f**, 30 dph and **g**, 50 dph) formation

In the wild, larvae feed on different types of microalgae, ciliates, copepods, mollusk eggs, polychaetes, larval decapods, and other marine fish larvae (Fortier and Harries 1989), visually (most of marine fish larvae are visual feeders) selecting the prey of the right size, slow movement (motion is a needed stimuli for many marine fish larvae to elicit a feeding response), and with a good nutritional profile, especially in terms of EFA, that cover larval requirements. For most marine fish larvae, copepods are the main live prey in the wild with the best nutritional profile (May 1974; Last 1979), but in aquaculture copepods have not been frequently used, and other prey such as rotifers (*Brachionus* sp., 50–250 m length) and nauplii (*Artemia* spp., 200–500 m) are widely used. Furthermore, in order to keep water quality conditions, live prey enriched, good visual conditions for the larvae, and a proper bacteriological environment, different microalgae (genus *Tetraselmis*, *Isochrysis*, *Nannochloropsis*, etc.) are added to the rearing tanks at least during the period of rotifer feeding. These live prey are added to the rearing tanks in sufficient amounts to be captured by the larvae in increasing numbers[1] per day. Depending on the marine fish species, a precise protocol for larval feeding is adopted.

Rotifers (*Brachionus* sp.) are commercially available and can be cultured in sufficient numbers to satisfy the needs of a marine fish hatchery (Dhert 1996; Lubzens and Zmora 2003; Dhont et al. 2013). They can be cultured with live algae, yeast, or algal paste to simplify the process and decrease the production costs. However, their composition in EFA cannot cover nutritional requirements of marine fish larvae and lack the most important fatty acids DHA, EPA, and ARA. Rotifers lack the ability to elongate or desaturate shorter chain fatty acids, and they must be enriched, to satisfy these EFA requirements, before they are fed to marine fish larvae (Sargent et al. 1999). Commercially available enrichment products are used to feed the rotifers, retaining adequate levels of EFA for several hours if kept cool. The process of enrichment is also known as bioencapsulation, and it is used not only to enrich live prey in fatty acids but in various kinds of nutrients. Several commercial and/or experimental enrichment diets have been used and include unicellular algae, emulsions, liposomes, and microencapsulated diets. The live prey are passive filter feeders and thus incorporate the enrichment/bioencapsulation products in their digestive tract, generally in the form of triacylglycerols, acting as live vehicles.

The quantity of rotifers in the larval rearing tanks must be constantly monitored to ensure that they are ingested by the larvae and that their density and nutritional quality is enough to ensure a good larval growth (Fig. 5.2).

The anostracan brine shrimp *Artemia* is the most widely used form of live food to culture larvae in the world. *Artemia* is harvested from natural hypersaline lakes as dormant cysts, which are easily collected, stored, and marketed. The cysts can be stored for long periods of time, and the rate of hatching can be predicted, making *Artemia* an attractive live food for the culture of many marine species. When the cysts are introduced into 28 °C saline water, they hatch within 24 h, and nauplii can be collected (Dhert 1996; Dhont et al. 2013). They are fed to the larvae after the

[1] It is estimated that a gilthead seabream larvae consumes apx 70 rotifers at first feeding, and this number is increased to 700 per day at 25 days post-hatching (Papandroulakis 2000).

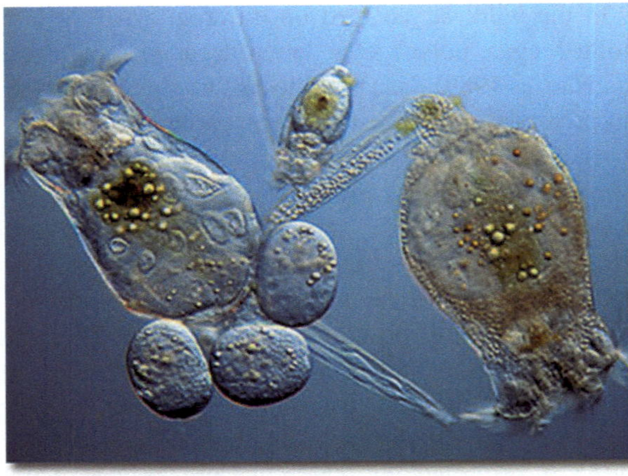

Fig. 5.2 Rotifer Brachionus sp. (www.gettyimages.com)

Fig. 5.3 Development of brine shrimp *Artemia* sp. from cysts to newly hatched nauplii, metanauplii, and adult stage. (www.zootecniadomestica.com)

rotifer feeding phase and up to the transition to an artificial diet. Some larger marine fish larvae (i.e., European sea bass, *Dicentrarchus labrax*, or Senegalese sole, *Solea senegalensis*) can be fed directly on *Artemia* nauplii at first feeding. Like rotifers they have insufficient levels of DHA, EPA, and ARA and need to be enriched as well, and same or similar products used for rotifer enrichment can also be used for *Artemia*.

As in the case of rotifers, the quantity of *Artemia* nauplii or enriched metanauplii in the larval rearing tanks must be constantly monitored to ensure that they are ingested by the larvae and that their density and nutritional quality are enough to ensure a good larval growth (Fig. 5.3).

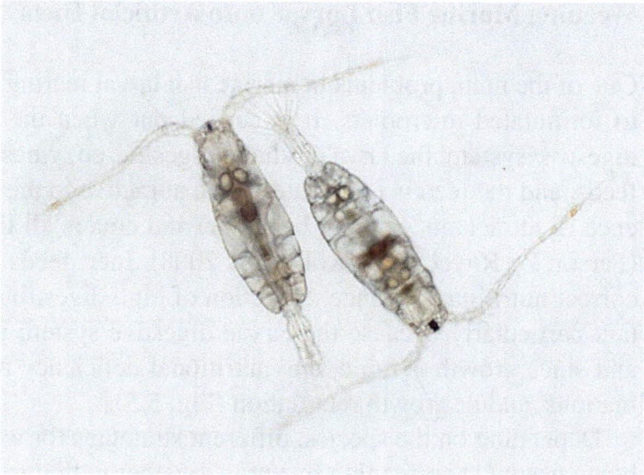

Fig. 5.4 Image of a Calanoid (planktonic) copepod *Acartia* sp. (www.CFeed.no)

Copepods are a major component of the marine zooplankton community and in the wild constitute a major link in the nutrient pathway from primary producers (microalgae) to marine fish larvae. Marine fish larvae eat copepod nauplii, whereas juvenile fish consume adults, being their role in the marine trophic system essential to the survival of many marine fish species. Their use for feeding marine fish larvae improves their growth and survival (Shields et al. 1999), contributes to reduce malpigmentation and impaired eye migration in flatfishes (Hamre et al. 2007), reduces stress, and increases larval fitness (Oie et al. 2015; Vanacor-Barroso et al. 2017). They are able to synthesize EFA, maintain an appropriate DHA/EPA ratio, and most of these EFA are present in copepods as phospholipids that can be more easily digested and used by the larvae. Apart from fatty acids and phospholipids, copepods are rich in astaxanthin (Dhont et al. 2013), a precursor of vitamin A. Astaxanthin working together with vitamin E suppress lipid peroxidation, being very important to fish with high levels of omega-3 polyunsaturated fatty acids (Bell et al. 2000). Although copepods are the preferred prey of wild marine fish larvae, their use in marine fish hatcheries remains sporadic (Ajiboye et al. 2011). Their attributes, especially in terms of nutritional composition, have increased the interest in large-scale culture, and recent literature has discussed culturing techniques and the use of copepods as live prey in marine fish aquaculture (see Franco et al. 2017). Copepod life cycles are long (Stottrup, 2003), with different phases (nauplii, copepodites, adults) that need to be kept in separated tanks due to adult cannibalism on eggs and nauplii (Drillet et al. 2014; Gallucci and Olafsson 2007). The culture methods are species-specific; they cannot be cultured using any other diet than live microalgae and require more space, equipment, and time than for culturing rotifers or *Artemia*; and the number of nauplii produced sometimes is not enough to cover industrial needs (Lee et al. 2005). Recently some private companies (CFeed AS) also offer copepod eggs, easy to hatch, and produce nauplii for commercial hatcheries that can be fed on microalgae in case that copepodites or adults are needed (Fig. 5.4).

Weaning Marine Fish Larvae onto Artificial Diets

One of the main problems in marine fish larval rearing is the transfer from live prey to formulated microdiets. It is carried out when the larva has a fully developed digestive system, the larva produces digestive enzymes for a better utilization of dry feeds, and the feed is formulated to be attractive to the fish (color, shape, size, presence of attractants, sinking behavior) and covers all their nutritional requirements (Person Le Ruyet 1989; Kolkovski 2008). Inert feeds must be processed to have a correct nutritional balance. Selection of high digestible components is very important particularly because the larvae digestive system is not completely functional, and since growth is rapid, any nutritional deficiency may cause disorders, malformations, and/or growth retardation (Fig. 5.5).

Depending on the species, different strategies for weaning marine fish larvae are being used: (1) co-feeding live prey together with microparticulate diets increasing the amount of feed given at the same time that *Artemia* nauplii supply is being reduced or (2) using a direct/drastic change from *Artemia* to microdiets. Nowadays commercial weaning diets are formulated using different attractants (mostly free amino acids), hydrolyzed proteins and peptides ready to be digested by the larvae, immune-stimulants (pre- and probiotics and other products) to protect digestive and immune systems, and other additives to avoid skeletal deformities. Although weaning is still a bottleneck, a better knowledge of nutrient requirements, digestive function of marine fish larvae and a wider selection of ingredients have increased the survival rate during this phase.

For organic aquaculture feed, the most salient issue is the existing bottleneck in supply of certified organic feed. Global demand for certified organic feed ingredients

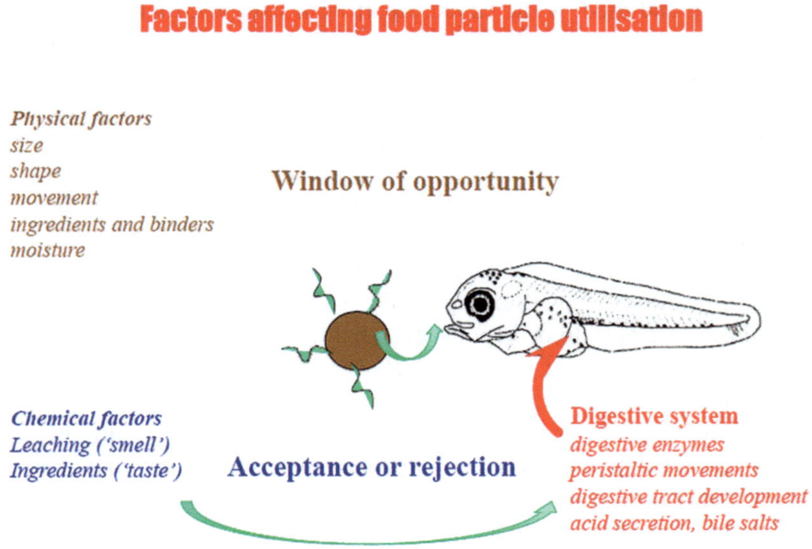

Fig. 5.5 Factors affecting feed utilization by marine fish larvae. (Kolkovski 2008)

for aquaculture and agriculture far outstrips supply, resulting in very high prices and consequently, high production costs. In a country with only one or a few organic aquaculture farms, the initiation of organic agriculture feed projects and the establishment of the first local organic aquaculture feed mill are a challenging process, requiring high levels of commitment by and cooperation between different sectors (e.g., aquaculture, agriculture, feed production). In many countries, existing feed mill operators hesitate to undertake the part-time production of relatively low amounts of feed due to the stringent requirements in preparing machines between runs of organic and nonorganic feed to avoid contamination.

As a consequence, if weaning diets for marine fish tend to be expensive, for organic aquaculture, the use of certified organic ingredients will increase the cost enormously, although recently new ingredients are considered acceptable in organic feeds (see Table 8.1, Chap. 8) and a reduction in the weaning feed prices is envisaged.

Freshwater Fish Larvae

Freshwater fish produce significantly larger and fewer eggs than marine fish, a difference that is not attributable to differences in body size (Elgar 1990). Freshwater fish larvae, compared to marine fish larvae, have a higher weight at hatch, have shorter larval stage durations, and have lower metabolic requirements, although growth rates and growth efficiencies are similar (Houde 1994). They also have lower probability of episodic mortalities and lower mortality due to starvation (mostly because marine fish larvae are smaller and have higher metabolic demands and higher ingestion requirements), and their mean survivorship is 44 times higher than that of marine fish larvae. The fatty acid requirements of freshwater fish larvae is different to that of marine fish; they need a certain proportion of LA and ALA to achieve maximal growth, whereas marine fish need ARA, EPA, and DHA derived from the elongation and desaturation of these two precursors (Tocher 2010). The ability of freshwater fish to elongate and desaturate dietary provided LA and ALA into long chain fatty acids of both n-3 and n-6 families is similar to that of birds and mammals. Most of them (salmonids, cyprinids, silurids, etc.) also have a functional stomach before yolk sac absorption that allows the larvae to easily adapt to a dry compound diet (Dabrowski 1984). Thus, in most of the freshwater farms, no live food is used, and the larvae are directly fed artificial diets formulated to cover their nutritional needs and ensure maximal growth and survival.

The only exceptions are several species of percids (i.e., pike perch, *Sander lucioperca*) that need to be fed on live prey, similar to marine fish larvae, mostly using enriched rotifers or small-sized *Artemia* nauplii, followed by enriched *Artemia* metanauplii (Lund et al. 2012) and common carp (*Cyprinus carpio*) that needs to be fed on non-enriched rotifers and copepod nauplii (Fig. 5.6).

Fig. 5.6 Newly hatched larva of salmon showing an opened mouth and pigmented eyes. (www.esacademic.com)

Crustacean Larval Development, Feeding, and Nutritional Requirements

Penaeid shrimp farming is dominated by only two species: Pacific or whiteleg shrimp, *Litopenaeus vannamei*, which is indigenous to the eastern Pacific (Mexico to Peru) and black tiger shrimp, *Penaeus monodon*, which is indigenous to the Indo-Pacific region (eastern Africa to Australia). Because of a number of characteristics considered favorable in farming conditions, *L. vannamei* was introduced into Asia and nowadays represents 80% of global shrimp aquaculture output (FAO 2018a), largely at the expense of *P. monodon*. Apart from these two species, a number of other species (*P. indicus*, *P. japonicus*, *P. sinensis*) are also still cultured in some regions.

Unlike fish larvae, crustacean larvae, and thus also penaeid shrimp larvae, grow discontinuously as they develop through a number of distinct larval stages. Each time, the larvae have to go through a molting or metamorphosis process in order to develop to the next stage. Each step in the development also involves considerable changes in terms of morphology, feeding behavior, physiology, etc.

Larval development of all farmed penaeid shrimp species is fairly similar. Females release eggs directly into the water as penaeids don't display brood care. After approximately 12–18 h (depending on incubation temperature), the eggs hatch out into a nauplius larvae, which is a nonfeeding stage that still thrives on the yolk reserves derived from the egg. In most penaeid species, there are five or six nauplii substages, called N1, N2, N3, etc. After about 48–72 h (depending on temperature), the nauplius develops into a zoea larva, which start exogenous feeding, and is primarily a herbivorous filter-feeder, feeding on microalgae and small particles. In hatchery operations, some ten algae species are commonly used. Typically, some smaller species such as the diatoms *Skeletonema or Chaetoceros* and/or *Isochrysis* are used initially. Later on larger cell size, microalgae like *Tetraselmis* may be introduced. A mixed algal diet seems to be preferred as it offers the best guarantee for a balanced nutrition (i.e., in terms of fatty acid profile). The zoea stage is also subdivided into three substages (Z1, Z2, and Z3). The Z3 stage then develops into a mysis larvae (three substages M1, M2 and M3) which become carnivorous. Feeding behavior also shifts from a passive filter-feeder to a visual predator. Since development in a population is never homogenous, zooplankton, normally *Artemia* nauplii, is usually introduced at the Z3 stage. In order to facilitate prey uptake, *Artemia* nauplii are often frozen or heat-killed in the 1st days of feeding. Later on live *Artemia* nauplii and sometimes also enriched metanauplii are fed. In some regions (i.e., Ecuador), rotifers (*Brachionus* spp.) are also used in the feeding schedule as an initial zooplankton, but this practice seems less common nowadays. When zooplankton is introduced, the amounts of algae fed are gradually reduced, but quite often a certain background level is maintained throughout the whole larval rearing phase as it stabilizes water quality. Apart from live food, most hatcheries also use different formulated feed as co-feed/partial live food replacement almost through the complete hatchery phase. Once the last mysis stage has passed metamorphosis to the postlarval (PL) stage, formulated feeds are almost exclusively used. In general, formulated feeds are rather easily accepted by penaeid shrimp larvae, and weaning doesn't cause any special problems (like it does for larvae of certain fish species). The complete larval development is also rather short and only takes 10–14 days to the postlarval stage. Because of the growing amounts of *Artemia* needed in that stage, the proportion of formulated feeds used become more important in order to reduce costs. In most places, postlarvae leave the hatchery to the grow-out ponds at PL10–PL12 stage (10–12 days after metamorphosis to PL). Because of disease issues, in certain places, it however seems to become popular to have an extended indoor nursery phase (e.g., in high-density biofloc systems) until the shrimp reach 1 g before they are transferred outdoors. In this way biosecurity is maximized, and the duration of the risky outdoor phase is reduced to the minimum (Fig. 5.7).

Nutritional Requirements and Feeding

Establishing exact nutritional requirements for shrimp larvae has been hampered by the lack of effective formulated feeds as a sole diet. Most research therefore has focused on testing optimal live food regimes, comparing different mixtures of algal species, alternative live foods, etc. (Wilkenfeld et al. 2009; Hadi Jamali et al. 2015). Quantitative requirements are mostly derived from studies in the postlarval or juvenile stage, considering that requirements during the larval stage are likely to be higher and digestive capabilities more limited. A recent review of shrimp nutritional requirements is provided by Christian Larbi Ayisi et al. (2017). As for fish, shrimp have specific requirements for protein: 25–35% crude protein might be sufficient for normal growth, but higher levels might be required at specific age/stage or in function of rearing conditions. Also, amino acid composition and protein digestibility are of utmost importance. The sulfur-containing methionine and lysine are usually the first limiting amino acids. For larvae and other young stages, partially hydrolyzed proteins may be advantageous in view of the limited digestive capabilities. Total lipid requirements of shrimp are overall lower than for fish. Nevertheless, a

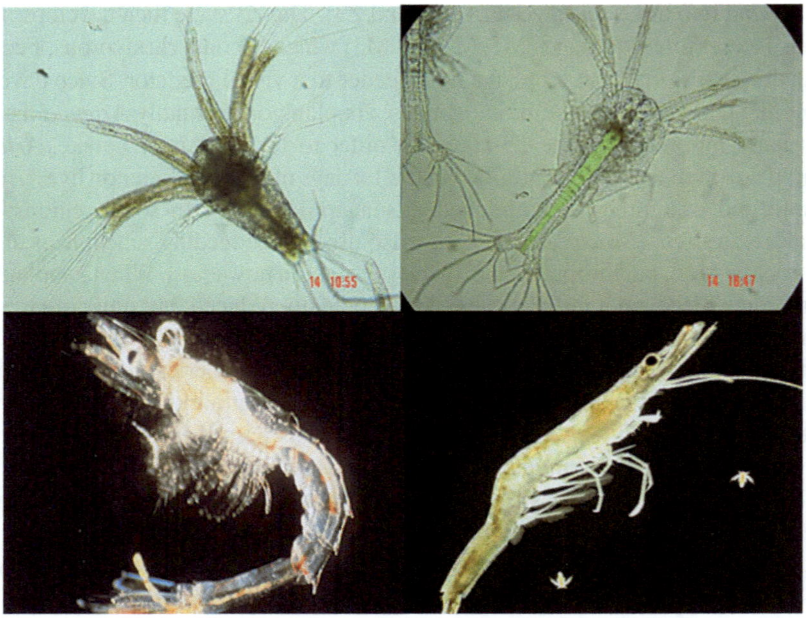

Fig. 5.7 Larval stages of penaeid shrimp: nauplius, zoea, mysis, and postlarva (with *Artemia* nauplii). (Courtesy: Roeland Wouters)

qualitative requirement for phospholipids, which are often included in formulated feeds at approximately 1.5%, has been documented for several species. Typically, shrimp also have a high requirement for cholesterol (0.5–1%). Furthermore, as for fish, a lot of research also focused on fatty acid and more specifically polyunsaturated fatty acid (PUFA) requirements of shrimp. It has been shown that feeding n-3 HUFA-rich diets results in improved survival, growth, and/or stress tolerance of several species. Bioencapsulation, also called *Artemia* enrichment or boosting, which is widely applied in marine fish hatcheries for enhancing the nutritional value of *Artemia* with essential fatty acids is much less common in shrimp larviculture, even though the benefit of feeding enriched *Artemia* nauplii to shrimp postlarvae has also been documented (Wouters et al. 2009; Mutti et al. 2017).

Essential nutrients may also be provided through live food supplements and formulated feeds, which are commonly used as co-feeds along the live food in shrimp hatcheries. In its simplest form, substitutes/partial replacements for live food used in shrimp hatcheries consist of dried algae or algal concentrates (pastes) and flakes made of *Artemia* or other organisms (i.e., krill). Complete formulated feeds exist in different forms: These include microbound diets, flakes, granulated feeds, microencapsulated feeds, and liquid feeds. Microbound feeds use a variety of binders to produce a small particle that is crumbled to the appropriate size. Flakes are commonly used in Asia and the Americas. To produce flakes, dietary ingredients are added to water to obtain a dense soup, with the resulting suspension pumped onto a steam drum dryer. Large flakes can be crushed and passed through an appropriate mesh screen immediately prior to use. Granulated feeds are produced using liquid binders and water sprayed onto the feed mix, resulting in granules with a raspberry-like structure. Microencapsulated feeds have an outer capsule coating that retains the ingredients inside. Liquid feeds are essentially slurry of particles in a suspension medium. Replacement of *Artemia* with formulated feeds has met variable success. The use of microbound feeds decreases survival or growth when fed at levels of 40–50% or higher. In controlled laboratory conditions, however, the results are generally better than in commercial hatcheries (Wouters et al. 2009) (Table 5.1).

Table 5.1 The effect of *Artemia* replacement

Species	Diet	*Artemia* replacement (%)	Larval stages	Result compared to *Artemia* control	References
Penaeus monodon	Crumbled experimental microbound diet	100	Z-PL	Similar survival but lower growth	Kanazawa et al. (1982)
P. monodon	Crumbled experimental microbound diet	100	Z-PL	Similar survival and growth	Kanazawa (1985)
P. monodon	Microencapsulated diet FRIPPAK	100	Z-PL	Similar survival and growth	Jones et al. (1989)
Litopenaeus vannamei	Microencapsulated diet	70–100	Z-PL	80% survival compared to 90% survival in live food control (commercial scale)	Jones et al. (1997)
L. vannamei	Crumbled experimental microbound diets	100%	M	Reduced development in experiment 1; reduced growth in experiment 2; similar survival, growth and development in experiment 3	D'Abramo et al. (2006)
L. vannamei	Crumbled microbound diets microfeast	25, 50, 75, 100	M-PL	Decreased growth rates at 50, 75 and 100% and decreased survival at 100%	Samocha et al. (1999)
Litopenaeus setiferus	Crumbled experimental microbound diets	40, 60, 100	Z-M	Decreased survival, growth, development, and stress resistance (but similar survival at 40 and 60% in the presence of algae)	Gallardo et al. (2002)

(continued)

Table 5.1 (continued)

Species	Diet	*Artemia* replacement (%)	Larval stages	Result compared to *Artemia* control	References
P. monodon	Microencapsulated diet FRIPPAK® FRESH	100	Z-PL	Increased survival, growth, and development (one single dose of live algae in zoea1)	Wouters et al. (2003b)
P. monodon	Crumbled microbound diet FRIPPAK® FLAKE	40, 100	PL	Lower survival, similar (100%), or improved (40%) growth	Wouters et al. (2003c)
L. vannamei	Crumbled microbound diet FRIPPAK® RW+	100	PL	Similar survival and growth in trial 1, lower survival and higher growth in trial 2 (98% survival in *Artemia* shell-free control)	Wouters et al. (2003a)
Farfantepenaeus aztecus	Liquid feeds EpifeedTM and LiqualifeTM	50, 100	M-PL	Decreased survival (except LiqualifeTM at 50%), growth and stress resistance	Robinson et al. (2005)
F. aztecus	Microbound diets ZeiglerTM E-Z larvae, ZeiglerTM Z-plus and E-Z *Artemia*	50, 100	M-PL	Decreased survival, growth and stress resistance	Robinson et al. (2005)
Fenneropenaeus chinensis	Crumbled experimental microbound diets	100	M-PL	Reduced growth and development	Wang and Mai (2006)

From Wouters et al. (2009)

Larval Culture Systems

Fish Larval Culture

For successful propagation of any marine fish species, larval rearing is a crucial phase and is often problematic. Unsuccessful methodologies for larval rearing of more-than-a-few species have delayed, for a long time, the diversification in marine finfish production. Today, the range of available hatchery techniques is diverse (Divanach and Kentouri 2000). The main classifications are based on the rearing density (intensive, semi-intensive, extensive) and the use of phytoplankton in the water (clear, green, pseudo-green) (Papandroulakis et al. 2002). All of them, however, share a common characteristic, i.e., the use of plankton (phyto- and zooplankton) during the period of first larval feeding.

In any applied method, there are three distinct phases during larval rearing: (i) egg hatching and autotrophic phase when larvae consume their yolk sac reserves, (ii) the heterotrophic phase when larvae are fed on zooplankton, and (iii) the weaning to artificial diets. During these phases larvae complete their transformation to juveniles. Juveniles usually remain in the hatchery, for pre-growing, until reaching 2–5 g in weight. In cases where on-growing is performed in open sea conditions, the pre-growing period is extended until individuals reach a weight of 10–30 g. This general scheme applies for both marine and freshwater larvae. A more detailed description of the applied techniques is presented in the following paragraphs.

Rearing Systems for Marine Larvae

Industrial production requires intensification in all the stages of the production process. In intensive hatcheries, larvae are reared at high densities under controlled conditions, and success is highly dependent on the level of knowledge of the larvae-specific biological needs. Growth and survival depend, to a great extent, on food availability and environmental conditions during rearing. Understanding the rate of food consumption and assimilation efficiency is of paramount importance to establish successful methodologies of larval rearing for aquaculture industry. The rate of consumption depends on food availability and the developmental stage of the organism.

The most commonly applied methods are (1) the "clear water" technique (Coves and Gasset 1993; Papandroulakis 2000), with no use of phytoplankton in the rearing medium, and (2) the so-called "pseudo-green water" technology (Papandroulakis et al. 2002) which is based on the frequent addition of phytoplankton and zooplankton in the larval rearing tanks, where phytoplankton is not produced, nor bloom, but its concentration remains constant by daily additions. (3) The "green water" technique that was based on the creation of optimum conditions for endogenous phytoplankton bloom of specific organisms in the larval tanks (Saroglia et al. 1989) is not further applied due to the difficult control of the tank environment. These methods are applied during the most critical segment of the rearing process, at the beginning of larval rearing (until the 20th–30th day post-hatching for the Sparidae, i.e., until the formation of the caudal fin), when the larvae are still extremely weak, sensitive to alterations in the rearing environment, easily stressed, and difficult to feed. After this period, the "clear water" methodology is applied. However, the biological requirements of only a few marine species are known enough to be reared at an industrial scale with intensive rearing methods.

Table 5.2 Main differences between hatchery techniques

Parameters	Techniques		
	Extensive	Mesocosm	Intensive[a]
Rearing enclosures	Ponds or bags	Tanks or bags	Tanks
Localization	Outdoor	Indoor[b]	Indoor
Rearing volume (m^3)	>100	30–100	<20
Rearing density (ind/l)	0.1–1	2–8	30–200
Food chain	Endogenous	Mixed	Exogenous
Infrastructures	Light	Medium	Sophisticated
Environment	Natural	Mixed	Controlled
Autonomy and autarky	High	Medium	Low to nil
Dependence on man and technique	Light	Medium	High to very high
Need for specific biological knowledge	Light	Medium	High to very high
Validity for new species	Very high	High	Medium to low

Divanach and Kentouri (2000)
[a]Integrate semi-intensive, intensive, and hyper-intensive techniques
[b]Sometimes outdoor (with bags) or semi-outdoor

In contrast, extensive hatcheries use large tanks or ponds under more natural conditions; larvae are reared at low densities, feeding on endogenous blooms of wild marine zooplankton. As a result, the probability of success is much higher than using the intensive approach, and the extensive method has been already employed for more than 20 fish species (Divanach and Kentouri 2000). Still, the challenge for industrial application of extensive methods is to increase their low productivity (Table 5.2).

The Mesocosm System

As an intermediate approach between the intensive and extensive method, semi-intensive techniques, like the so-called mesocosm technology (Divanach and Kentouri 2000), have been developed and are applied for the rearing of several species and can thus be considered as a semi-intensive technique of mass production. The actual form of the mesocosm technology was defined in the early 1990s after studying the originally applied models of extensive rearing (Grice and Reeve 1982; Divanach 1985; Lalli 1990). It requires indoor or semi-outdoor infrastructures. The most important characteristic is the size of the larval tanks, which should range between 20 and 60 m^3. Both natural and artificial conditions are combined during the rearing, thus making the method independent of any climatic and/or seasonal changes. There is a partial control of the light conditions (intensity and photophase) and a minimal control of the temperature. Two variants of the mesocosm exist according to the origin and quality of the food chain.

In the extensive variant, the food chain is basically endogenous and complemented with exogenous input when symptoms of overgrazing appear. In the intensive variant, the food chain is basically exogenous but exhibits a limited capacity of endogenous production, due to both the low density of larvae and the presence of phytoplankton in the environment. In both cases, the technology is characterized by a partial food autarky, and human error during rearing is less likely to have lethal effects on the larval population. Food, from both endogenous and exogenous origin, ensures a good matching of larval energy requirements, while avoiding any possibility of quality deficiencies or any risk of over grazing of the food chain. This partial autarky is very important for the effective application of the technology.

The initial egg density in the mesocosm ranges from 4 to 7 eggs/l, depending on species, and should never exceed 20 eggs/l. Tanks are filled with natural seawater filtered mechanically, and wild plankton is thus introduced in the system offering a capacity for endogenous production. Phytoplankton is added daily to maintain the green medium for a period of 2–4 weeks after hatching. Exogenously produced enriched rotifers, enriched *Artemia* sp. metanauplii and artificial diet are added when required. Application of the above procedure allows a zooplankton concentration higher than 1.5 and 0.1 individuals ml^{-1} for rotifers and *Artemia*, respectively, during the rearing period. In most cases, endogenously produced copepods (Harpacticoida, Tisbidae) are developed on tank walls contributing also as food items for the larvae.

A main difference between mesocosm and standard intensive methods, apart from the higher stocking density of the larvae and the smaller rearing volumes in the intensive method, is related to the food availability in terms of quality and quantity (Divanach and Kentouri 2000). Food quality differs between the two systems as, depending on the origin of water, a natural food chain can be developed during a mesocosm rearing that, in turn, can contribute to the diet of the larvae, significantly affecting the final results (Papandroulakis et al. 2005). Also larval growth/developmental rates differ between the two systems. In general, European sea bass reared in mesocosm have faster growth rates during the larva and early juvenile stages, with shorter time to and faster metamorphosis, less vertebrae deformities, and higher female to male ratio (Papadaki et al. 2005).

The mesocosm methodology results in high survival rates (about 55% from the initial eggs incubated after 50–70 days of rearing) and low percentage of individuals with developmental abnormalities (less than 2%), while, in general, larval growth performance is better than in the classical intensive systems. A technology, however, is only accepted for industrial application if it is proved economically feasible. The production cost per individual is comparative to standard methods or even lower (Papandroulakis et al. 2004).

Similar semi-intensive methods, like the above described, are also applied in different parts of the world, under different names such as "large volume rearing" (Prestinicola et al. 2013; Dhert et al. 1998) where the size of tanks, the rearing density, and the presence of wild plankton are critical factors of the process. Recent studies (Prestinicola et al. 2013) concluded that large volume rearing leads to a significant improvement of the morphological quality (i.e., lowered incidence of severe skeletal anomalies and meristic count variability) of gilthead seabream juveniles reared under semi-intensive conditions.

Mesocosm methodology may also result in alternate behavior during on growing. In particular, European sea bass coming from mesocosm systems, under some conditions (e.g., high temperature period), reacted more actively than the intensively reared group, as expressed by the observed displacements in the cage (Papandroulakis et al. 2011). Bearing in mind that the mesocosm conditions are more natural compared to the intensive ones, the individuals of mesocosm origin may be closer to the wild pattern than their counterparts from intensive rearing.

Crustacean Larval Culture

Hatchery Systems

In contrast to the limited number of species being farmed, hatchery systems being used for penaeid shrimp are quite divers and differ largely with geographical area, country, and even region within a country.

In a historical perspective, three types of hatcheries can be distinguished (Wickins and Lee 2002). In Southeast Asia, small-scale "backyard hatcheries" were quite common. These hatcheries use a low-technology, low-cost approach, with small tanks (less than 10 tons), and relatively low stocking densities. They are rather flexible and can adapt to market conditions (demand for PL) quickly. They often specialize in only one part of the life cycle (e.g., growing nauplii to PL) and purchase nauplii and microalgae from other small specialized companies. With this system survival rate can vary from 0% to 90%, but as operating costs are low, occasional failure is not a big problem.

Another type of shrimp hatchery technique is often referred to as the Eastern, Japanese/Taiwanese large tank technique. It was first developed in Japan (for *Penaeus japonicus*) and later on improved in Taiwan. It uses very large tanks (even up to 150 tons). Breeders are spawned within the larval rearing tanks. The water in the tanks is fertilized to stimulate algal blooms and zooplankton growth. Additional plankton (whether wild or cultured) may be added if natural productivity proves too low. This technique uses a kind of ecosystem approach, which could in part be compared with the mesocosm system described above for fish larvae. The main disadvantage is that this system is rather difficult to control, but survival can nevertheless be rather high (40% to PL). One could argue that this hatchery system in a number of aspects complies best with the philosophy of organic farming.

A last hatchery technique is referred to as the Western or Galveston (in Texas, USA, where it was first developed) technique. It uses relatively large tanks (10–30 tons). A high-tech approach is applied and everything is very much controlled. Water is treated effectively and exchanged at high rates. Spawning and hatching happens in separate tanks. Live food is cultured separately and densities in the larval rearing tanks managed well. Survival can be as high as 80%, but (because of disease issues) total failure also happens.

More recently, the difference between these types of hatcheries has become increasingly blurred, and overall there is a tendency for more controlled high-tech intensive-type hatchery techniques. Backyard hatcheries still do exist in certain regions, but they usually serve the supply of small-scale farms. Bigger companies quite often became vertically integrated and consist of a number of grow-out farms, one or more hatcheries, and sometimes their own feed mill or processing plant. In Latin America, for example, in Ecuador, big hatcheries have always been the trend.

More advanced techniques for water pretreatment (i.e., ozonization or UV treatment), disease diagnosis, and/or use of pure oxygen are becoming more and more adopted, especially in larger hatcheries. Recently, also application of intensive zero-exchange rearing systems based on RAS and biofloc technology is being tested for larval rearing (de Lourdes Cobo Barcia 2013; de Lorenzo et al. 2016). In the latter, a carbon source is added to stimulate growth of heterotrophic bacteria, trapping the nitrogenous waste accumulating in the system and converting it in microbial biomass. The microbial floc may then be consumed by the fish or shrimp and in this way improve feed efficiency of the system considerably. Ekasari and Bossier (2017) provided a review on the application of biofloc technology in aquaculture. Although biofloc systems are been tested for larval rearing of shrimp as mentioned above, they seem to be more readily adoptable to nursery and on-growing culture since larger floc may physically trap young larvae and are not efficiently ingested by the small larvae. Therefore, it seems to be necessary to steer the system to smaller floc size if intended for larval stages. Moreover, although RAS and biofloc technology aim to improve impact on the environment/feed efficiency, and in this sense correspond very well with the principles of organic production, they require high oxygenation and/or agitation of the water and thus have extra energy requirements, which is considered in the current EU regulation (Reg. EC 889/08, Article 25g) an acceptable compromise for organic hatchery and nursery but not allowed for the on-growing phase. A special type of hatchery exists in the more inland regions of the Mekong delta in Vietnam, which uses concentrated brine to make up seawater and apply low-tech recirculating systems to culture shrimp larvae in order to save on water consumption (Thanh et al. 2005).

Rearing Fish Larvae Under Organic Regulations

For Freshwater Fish

As indicated above the advantage of the freshwater fish larvae is that most of them already have pigmented eyes and an almost totally developed digestive system after hatching; this allow them to feed directly artificial diets instead of live prey after completion of yolk sac resorption. In the case of trout, yolk is depleted about 120 day degrees (14–20 days at 7 °C, Jokumsen and Svendsen 2010), and the fry can be moved to the tanks or ponds where the on-growing will take place. In the case of tilapia (Milstein and Lev 2004), most of the organic production is based on genetically selected lines that result in a high male proportion, and as in trout the larva can be fed directly on artificial diets. For other freshwater fish, like carp or catfish, similar methods are used, and once the larvae have hatched, they are maintained in the hatchery until yolk resorption and then transferred to the tanks or ponds, where the on-growing will take place. For most of the freshwater fish, the on-growing takes place in polyculture systems in ponds at low densities.

For organic culture of these freshwater fish, apart from being cultured at low densities in extensive or semi-intensive systems, (i) in the larval stage, the fish can be fed with conventional phytoplankton and zooplankton (Reg. EC 889/08, Article 25la); (ii) in the weaning stage, feed mixes are different from those for on-growing. It may also be technically possible to produce organic feeds for weaning considering the different nutritional requirements for freshwater larvae, and manufacturers would be interested to produce such feeds if there is sufficient demand.

For Marine Fish

As indicated previously in this chapter, marine fish larvae at hatching cannot see, detect, or ingest any artificial food, and only few days after hatching, when the eyes become pigmented, the gut is opened, and the larvae began to swim, they are able to follow, capture, and ingest small live prey (zooplankton). In conventional aquaculture, enriched rotifers and *Artemia* nauplii and metanauplii are often used for larval feeding. When the larvae are reared using a mesocosm system, they are kept at low densities, with seawater filtered only mechanically in order to allow the entrance of other zooplankton apart from the rotifers and *Artemia* nauplii administered by the producers; thus a natural food chain can be developed during a mesocosm larval rearing. Due to the similarities of this technique with those observed in the wild, this method has been evaluated by the EU Expert Group for Technical Advice on Organic Production (EGTOP) as the most adequate for organic production of marine larval fish (EGTOP 2014). However, the use of fatty acid enrichment, which is essential for a normal larval development, remains an unresolved problem because most of commercial enrichment products are made up with synthetic antioxidants and emulsifiers that do not comply with organic standards. Similarly, phytoplankton needs to be cultured on a medium based on commercial nutrient solutions of macro- and micronutrients, silicates, and vitamins in easily soluble, mineral form. All these substances would need to be approved and included in the specific positive lists of the EU organic regulation in order to allow the organic certification of phytoplankton. Due to these two handicaps, the use of nonorganic zooplankton (enriched or bioencapsulated) and phytoplankton is, at present, allowed by EU organic regulation (Reg. EC 889/08, Article 25la).

For Penaeid Shrimp

As mentioned earlier, because of some advantageous characteristics, *Litopenaeus vannamei* has become the preferred species for farming, also in Asia. *L. vannamei* is however not indigenous to Asia, and therefore, according to regulations for organic production, farming in Asia would have to be restricted to *Penaeus monodon* (or another indigenous species).

P. monodon has however the disadvantage that domestication and controlled breeding is far less developed as compared to *L. vannamei*. Where for *L. vannamei* special brood stock companies providing domesticated and selected breeding lines, which are traded internationally, exist, this is far less the case for *P. monodon*. For *P. monodon* it is still very common to resort to wild caught animals to serve as brood stock (Debnath et al. 2016). However, in the medium-long term, also for *P. monodon,* the target is to provide domesticated and selected breeding lines as organic brood stock although periodically wild caught animals will be used in order to avoid inbreeding.

Moreover, where maturation and spawning conditions are far better mastered for *L. vannamei*, in part also because with continuing domestication natural maturation and spawning seems to improve, reproduction is still less understood and controlled for *P. monodon*. In this sense, especially for *P. monodon*, it is also still a common practice to apply eyestalk ablation (removal of one of the eyes and eyestalks) in order to speed up maturation and spawning (Debnath et al. 2016). Eyestalk ablation of the females removes the X-organ-sinus gland complex that is the source of one of the reproductive inhibitory substances, and as a result maturation, spawning success and fecundity are enhanced. This practice is however questionable in terms of animal welfare and therefore not allowed in organic farming.

From the above, it is clear that, with the current knowledge and state of technology, the prerequisite to use *P. monodon* for organic shrimp farming in Asia, due to difficulties in terms of brood stock sourcing and control of reproduction, makes production of organically grown postlarvae very difficult.

As explained earlier, microalgae are an important live food for the early shrimp larval stages. Reliable production of microalgae however largely relies on the use of inorganic nutrients (nitrates, phosphates, silicates), which can be dosed precisely in the optimal ratio for the different algal species. As mentioned earlier, the use of these commercial inorganic nutrients is still an unresolved problem for the organic regulation that would prefer organic fertilizers as nutrient source. It is however not sure if the use of organic fertilizers would allow to produce microalgae reliably on a large scale. Replacement of autotrophically grown algae with heterotrophically grown algae or other heterotrophic marine protists, such as Thraustochytrids (e.g., *Schizochytrium* spp.), could potentially solve this issue. This however needs further research.

Although less common than for fish larvae, the *Artemia* nauplii used as live food for later shrimp larval stages are also some times enriched with special enrichment products in order to boost the level of PUFA and other nutrients. Also, in this case enrichment products based on yeasts, heterotrophically grown algae or thraustochytrids may provide a solution. Sprague et al. (2017) provide a good review on the use of microbial oils as replacement for fish oil in aquaculture. As mentioned for fish larvae, these commercial enrichment products may also potentially contain several synthetic ingredients such as emulsifiers, ethyl esters of fatty acids, synthetic carotenoids, antioxidants, and vitamins, which would be in conflict with organic requirements.

Similarly, formulated feeds used as supplement or partial replacement of the live food may contain several ingredients or substances that are not allowed under organic regulations. Because of the high digestibility and optimal composition in terms of amino and fatty acids, formulated feeds moreover may contain a relatively large proportion of protein and lipid from marine animal origin (fish meal and fish oil). In view of sustainability issues regarding fish meal and fish oil use, organic farming regulation however recommend minimal use of these ingredients. Possible alternatives for fish oil were already mentioned earlier. In the last decades, considerable research effort was directed to replace FM by alternative protein sources, including plant proteins (soybean, canola meal, etc.), microbial proteins (so-called single cell proteins), and alternative animal proteins such as meat and bone meal, poultry by-product meal, and more recently also more innovative sources such as mealworm (Panini et al. 2017) and black soldier fly meal (Cummins et al. 2017). It seems most alternative protein sources can successfully replace part of the fish meal, but to date none can replace FM 100% without negative effects on either survival or growth performance.

Conclusions

Organic agriculture (and by extension organic aquaculture) is still a young sector but one of the most dynamic in Europe. Recent advances on EU regulations have been made by EGTOP regarding the authorization of products, substances, and techniques for use in organic aquaculture, but there are several subjects, especially related to marine larval fish production that need further attention and solutions. Most of the problems are a consequence of the use of phytoplankton (microalgae) and zooplankton for larval feeding, instead of commercially balanced feeds, formulated using ingredients approved by organic production rules and produced under these principles, like those used for organic freshwater larvae.

Thus, further research and/or solutions are needed for the following topics:

1. Domestication and brood stock management. This is especially urgent for *P. monodon* in support of organic shrimp farming in Asia.
2. The use of water-soluble fertilizers for phytoplankton, such as pure inorganic salts, needs to be reviewed and considered.
3. Research on the use of organic fertilizers that can be combined with microalgae production.

4. In the case of rotifer and *Artemia* enrichment (bioencapsulation), most of the commercially available products contain oils, emulsifiers, and supplements that do not comply with organic standards. In this case emulsifiers like lecithin derived from organic raw material and vitamins A, D, and E derived from vegetable oils/products are permitted by EU regulations (EU Regulation 2016/673). Also new enrichment products, rich in omega-3 fatty acids (mostly DHA), derived from the culture of heterotrophic microalgae such as *Schyzochytrium* sp. can be considered as substitutes for emulsions.
5. Research in the enrichment of live prey using only microalgae or combined with other organic products is also a need.
6. Effective formulated diets (including brood stock and weaning diets) that comply with organic regulations: reduced levels of fish meal and fish oil, free of synthetic ingredients. This includes research into alternative protein and lipid sources, alternatives for synthetic emulsifiers, vitamins, pigments, antioxidants, etc.
7. The use of essential amino acids produced through fermentation should be reviewed and considered, as for the histidine for salmon (Reg. EC 889/08, Article 25k).

References

Ajiboye OO, Yakubu AF, Adams TE, Olaji ED, Nwogu NA (2011) A review of the use of copepods in marine fish larviculture. Rev Fish Biol Fish 21:225–246

Ayisi CL, Hua X, Apraku A, Afriyie G, Kyei BA (2017) Recent studies toward the development of practical diets for shrimp and their nutritional requirements. Hayati J Biosci 24:109–117

Bell JG, McEvoy LA, Tocher DR, Sargent J (2000) Depletion of α-tocopherol and astaxanthin in Atlantic salmon (*Salmo salar*) affects autoxidative defense and fatty acid metabolism. J Nutr 130:1800–1808

Coves D Gasset E (1993) Gilthead seabream (*Sparus aurata*) intensive larva rearing in closed systems. In: Proceeding of the World Aquaculture 93. Special Publication European Aquaculture Society no. 19:342

Cummins VC, Rawles SD, Thompson KR, Velasquez A, Kobayashi Y, Hager J, Webster CD (2017) Evaluation of black soldier fly (*Hermetia illucens*) larvae meal as partial or total replacement of marine fish meal in practical diets for Pacific white shrimp (Litopenaeus vannamei). Aquaculture 473:337–344

D'Abramo LR, Perez EI, Sangha R, Puello-Cruz A (2006) Successful culture of larvae of *Litopenaeus vannamei* fed a microbound diet exclusively from either stage PZ2 or M1 to PL1. Aquaculture 261:1356–1362

Dabrowski K (1984) The feeding of fish larvae: present "state of the art" and perspectives. Reprod Nutr Dev 24:807–833

de Lorenzo MA, Poli MA, Candia EWS, Schleder DD, Rodrigues MS, Vieira FD (2016) Hatchery performance of the pacific white shrimp in biofloc system using different stocking densities. Aquacult Eng 75:46–50

de Lourdes Cobo Barcia M (2013) Intensification of white shrimp *Litopenaeus vannamei* (Boone) larviculture. PhD thesis, Ghent University, 200 p

Debnath P, Khan SH, Karim M, Belton B, Mohan CV, Phillips M (2016) Review of the history, status and prospects of the black tiger shrimp (Penaeus monodon) hatchery sector in Bangladesh. Rev Aquac (2016) 8:301–313

Dhert P (1996) Roifers. In: Lavens P, Sorgeloos P (eds) Manual on the production and use of live food for aquaculture, FAO Fish Tech Paper 361. FAO, Roma, pp 49–78

Dhert P, Divanach P, Kentouri M, Sorgeloos P (1998) Rearing techniques of difficult marine fish larvae. World Aquac 29(1):48–55

Dhont J, Dierckens K, Stottrup J, Van Stappen G, Wille M, Sorgeloos P (2013) Rotifers, Artemia and copepods as live feeds for fish larvae in aquaculture. In: Advances in aquaculture hatchery technology. Woodhead Publishing Limited, pp 157–202

Divanach P (1985) Contribution a la connaissance de la biologie et de l' élevage de 6 sparides Méditerranéens: *Sparus aurata*, *Diplodus sargus*, *Diplodus vulgaris*, *Diplodus annularis*, *Lithognathus mormyrus*, *Puntazzo puntazzo* (*Poissons téléostéens*). Thèse de doctorat dès Sciences, Université de Sciences et Techniques du Languedoc, Montpellier, 479 p

Divanach P, Kentouri M (2000) Hatchery techniques for specific diversification in Mediterranean finfish larviculture. Cah Options Mediterr 47:75–87

Drillet G, Maguet R, Mahjoub M-S, Roullier F, Fielding MJ (2014) Egg cannibalism in *Acartia tonsa*: effects of stocking density, algal concentration, and egg availability. Aquac Int 22:1295–1306

EGTOP (2014) Final report on aquaculture (part B). https://ec.europa.eu/agriculture/organic/eu-policy/expert-advice/documents/final-reports_en

Ekasari J, Bossier P (2017) Biofloc technology application in aquaculture to support sustainable development goals. Microb Biotechnol 10(5):1012–1016

Elgar MA (1990) Evolutionary compromise between few large and many small eggs: comparative evidence in teleost fish. Oikos 59:283–287

Estévez A, Ishikawa M, Kanazawa A (1997) Effects of arachidonic acid on pigmentation and fatty acid composition of Japanese flounder *Paralichthys olivaceus* (Temmink and Shlegel). Aquac Res 28:279–289

FAO (2018a) Fisheries statistical collections. Global Aquaculture Production. Online query: http://www.fao.org/fishery/statistics/global-aquaculture-production/en

FAO (2018b) The State of World Fisheries and Aquaculture 2018 - Meeting the sustainable development goals. Rome. Licence: CC BY-NC-SA 3.0 IGO

Fortier L, Harries RP (1989) Optimal foraging density-dependent competition in marine fish larvae. Mar Ecol Prog Ser 51:19–33

Franco SC, Augustin CB, Geffen A, Dinis MT (2017) Growth, egg production and hatching success of *Acartia tonsa* cultured at high densities. Aquaculture 468:569–578

Gallardo PP, Pedroza-Islas R, García-Galano T, Pascual C, Rosal C, Sánchez A, Gaxiola G (2002) Replacement of live food with a microbound diet in feeding *Litopenaeus setiferus* (Burkenroad) larvae. Aquac Res 33:681–691

Galluci F, Olafsson E (2007) Cannibalistic behaviour of rock-pool copepods: an experimental approach for space, food and kinship. J Exp Mar Biol Ecol 342:325–331

Govoni JJ, Boehlert GW, Watanabe Y (1986) The physiology of digestion in fish larvae. Environ Biol Fish 16:59–77

Grice GD, Reeves MR (eds) (1982) Marine mesocosms, biological and chemical research in experimental ecosystems. Springer-Verlag, New York

Hamre K, Holen E, Moren M (2007) Pigmentation and eye migration in Atlantic halibut (*Hippoglossus hippoglossus* L.) larvae: new findings and hypotheses. Aquac Nutr 13:65–80

Houde ED (1994) Differences between marine and freshwater fish larvae: implications for recruitment. ICES J Mar Sci 51:91–97

Jamali H, Ahmadifard N, Abdollahi D (2015) Evaluation of growth, survival and body composition of larval white shrimp (*Litopenaeus vannamei)* fed the combination of three types of algae. Int Aquat Res 7(2):115–122

Jokumsen A, Svendsen L (2010) Farming of freshwater rainbow trout in Denmark. DTU Aqua Report no. 219-2010. ISBN 978-87-7481-114-5

Jones DA, Amjad S, Chitravadivelu K (1989) Comparison of artificial feeds used in penaeid shrimp hatcheries. In: Nour AM (ed). Proceedings of the 3rd Egyptian-British Conference on Animal, Fish and Poultry Production. University College of North Wales, Bangor, UK, pp 15–20

Jones DA, Yule AB, Holland DL (1997) Larval nutrition. In: D'Abramo LR, Conklin DE, Akiyama DM (eds) Crustacean nutrition. Advances in world aquaculture, vol 6. World Aquaculture Society, Baton Rouge, pp 353–389

Kanazawa A, Teshima S, Sasada H, Rahman SA (1982) Culture of prawn larvae with micro-particulate diets. Bull Jpn Soc Sci Fish 48:195–199

Kanazawa A (1985) Microparticulate diets. In: Yone Y (ed) Fish nutrition and diets. Koseisha-Koseikatu, Tokyo, pp 99–110

Kolkovski S (2008) Advances in marine fish larvae diets. In: Avances en Nutricion Acuicola IX. Univ. Autónoma de Nuevo León, Monterrey, pp 20–45

Koven W, Barr Y, Lutzky S, Ben-Atia I, Weiss R, Harel M, Behrens P, Tandler A (2001) The effect of dietary arachidonic acid (20:4n-6) on growth, survival and resistance to handling stress in gilthead seabream (*Sparus aurata*) larvae. Aquaculture 193:107–122

Lalli CM (1990) Enclosed experimental marine ecosystems: a review and recommendations, Coastal and marine studies. Springer-Verlag, New York

Last JM (1979) The food of larval turbot Scohpthalmus maximus L. from the west central North Sea. J Cons Int Explor Mer 38:208–313

Lee CS, O'Bryen PJ, Marcus NH (2005) Copepods in aquaculture. Blackwell Publishing, Ames. 269 pages

Lubzens E, Zmora O (2003) Production and nutritional value of rotifers. In: Stottrup JG, McEvoy LA (eds) Live feeds in marine aquaculture. Blackwell Science Publishing Ltd, Oxford, pp 17–64

Lund I, Skov PV, Hansen BW (2012) Dietary supplementation of essential fatty acids in larval pikeperch (*Sander lucioperca*); short and long term effects on stress tolerance and metabolic physiology. Comp Biochem Physiol A 162:340–348

May RC (1974) Larval mortality in marine fishes and the critical period concept. In: Blaxter JHS (ed) The early life history of fish. Springer-Verlag, New York, pp 1–19

Milstein A, Lev O (2004) Organic tilapia culture in Israel. In: Bureau of fisheries and Aquatic Resources (BFAR) & American tilapia Association (ATA) (ed) 6th international symposium on Tilapia in aquaculture, ISTA 6. ATA. Philippines, Manila

Mourente G, Tocher DR (1992) Effect of weaning onto a pelleted diet on docosahexaenoic (22:6n-3) levels in brain of developing turbot (*Scophthalmus maximus* L.). Aquaculture 105:363–377

Mutti DW, Ballester ELC, Martino RC, Wasielesky W Jr, Cavalli RO (2017) Feeding n-3 HUFA enriched *Artemia* to the larvae of the pink shrimp *Farfantepenaeus paulensis* increases stress tolerance and subsequent growth. Lat Am J Aquat Res 45(1):18–24

Oie G, Galloway T, Soroy M et al (2015) Effect of cultivated copepods (*Acartia tonsa*) in first-feeding of Atlantic cod (*Gadus morhua*) and Ballan wrasse (*Labrus bergylta*) larvae. Aquac Nutr 23:3–17. https://doi.org/10.1111/anu.12352

Panini RL, Freitas LEL, Guimaraes AM, Rios C, da Silva MFO, Vieira FN, Fracalossi DM, Samuels RI, Prudencio ES, Silva CP, Amboni RDMC (2017) Potential use of mealworms as an alternative protein source for pacific white shrimp: digestibility and performance. Aquaculture 473:115–120

Papadaki M, Piferrer F, Zanuy S, Maingot E, Divanach P, Mylonas CC (2005) Growth, sex differentiation and gonad and plasma levels of sex steroids in male- and female dominant populations of *Dicentrarchus labrax* obtained through repeated size grading. J Fish Biol 66:938–956

Papandroulakis N (2000) Influence of the rearing conditions on the development and food consumption of sea bream (*Sparus aurata*) during the early developmental stages. Mathematical simulations. Biology dept, School of Positive Sciences, University of Crete, Heraklion, 196 p. (in Greek, with English abstract)

Papandroulakis N, Divanach P, Anastasiadis P, Kentouri M (2002) The pseudo-green water technique for intensive rearing of sea bream (*Sparus aurata*) larvae. Aquac Int 9:205–216

Papandroulakis N, Kentouri M, Maingot E, Divanach P (2004) Mesocosm a reliable technology for larval rearing. Application with *Diplodus puntazzo* and *Diplodus sargus sargus*. Aquac Int 12:345–355

Papandroulakis N, Mylonas CC, Maingot E, Divanach P (2005) First results of greater amberjack (*Seriola dumerili*) larval rearing in mesocosm. Aquaculture 250:155–161

Papandroulakis N, Lika K, Kristiansen T, Oppedal F, Divanach P, Pavlidis M (2011) Behaviour of European sea bass, *Dicentrarchus labrax* L., in cages – impact of early life rearing conditions and management. Aquac Res. https://doi.org/10.1111/are.12103

Person-Le Ruyet J (1989) Early weaning of marine fish larvae onto microdiets: constraints and perspectives. Actes de Colloque 9:625–642

Prestinicola L, Boglione C, Makridis P, Spano A, Rimatori V, Palamara E, Scardi M, Cataudella S (2013) Environmental conditioning of skeletal anomalies typology and frequency in gilthead seabream (*Sparus aurata* L., 1758) juveniles. PLoS One 8(2):e55736. https://doi.org/10.1371/journal.pone.0055736

Robinson CR, Samocha TM, Fox JM, Gandy RL, McKee DA (2005) The use of inert artificial commercial food sources as replacements of traditional live food items in the culture of larval shrimp, *Farfantepenaeus aztecus*. Aquaculture 245:135–147

Sargent J, Bell JG, McEvoy L, Tocher DR, Estévez A (1999) Recent developments in the essential fatty acid nutrition of fish. Aquaculture 177:191–199

Saroglia M, Ingle E, Marino G, Del Vecchio ML, Porrello S (1989) Produzione commerciale di avannotti di spigola in Italia. Contributo della tecnologia acque verdi. Nova Thalassia 10:309–317

Samocha TM, Matsumoto T, Jones ER, Torano M (1999) Use of artificial diets to reduce Artemia nauplii requirements for production of *Litopenaeus vannamei* postlarvae. The Israeli Journal of Aquaculture – Bamidgeh 51:157–168

Shields RJ, Bell JG, Luizi FS, Gara B, Bromage N, Sargent JR (1999) Natural copepods are superior to enriched Artemia nauplii as feed for halibut larvae (*Hippoglossus hippoglossus*) in terms of survival, pigmentation and retinal morphology: relation to dietary essential fatty acids. J Nutr 129:1186–1194

Sprague M, Betancor MB, Tocher DR (2017) Microbial and genetically engineered oils as replacements for fish oil in aquaculture feeds. Biotechnol Lett 39:1599–1609

Støttrup JG (2003) Production and nutritional value of copepods. In: Støttrup JG, McEvoy LA (eds) Live feeds in marine aquaculture. Blackwell Science, Oxford, pp 145–205

Thanh T, Trong Nghia T, Wille M, Sorgeloos P (2005) Recirculation systems: Sustainable alternatives for backyard shrimp hatcheries in Asia? Aquac Asia Mag. July–September, pp 32–33

Tocher DR (2010) Fatty acid requirements in ontogeny of marine and freshwater fish. Aquac Res 41:717–732

Vanacor-Barroso M, Carvalho CVAD, Antoniassi R, Ronzani-Cerqueira V (2017) The copepod *Acartia tonsa* as live feed for fat snook (*Centropomus parallelus*) larvae from notochord flexion to advanced metamorphosis. Lat Am J Aquat Res 45:159–166

Villalta M, Estevez A, Bransden MP (2005) Arachidonic acid enriched live prey induces albinism in Senegalese sole (*Solea senegalensis*) larvae. Aquaculture 245:193–209

Wang X, Mai K (2006) A successful microbound diet for the larval culture of Chinese shrimp *Fenneropenaeus chinensis*. J Ocean Univ China 4:267–271

Wickins JF, Lee DO'C (2002) Crustacean farming ranching and culture, 2nd edn. Blackwell Science Ltd., Oxford. 445 pp

Wilkenfeld JS, Lawrence AL, Kuban KD (2009) Survival, metamorphosis and growth of penaeid shrimp larvae reared on a variety of algal and animal foods. J World Aquacult Soc 15(1–4):31–49

Wouters R, Van Horenbeeck T (2003a) Larval shrimp feeds: current status. In: Jory DE (ed). Responsible Aquaculture for a Secure Future: Proceedings of a Special Session on Shrimp Farming. World Aquaculture 2003, Recife, Brazil. The World Aquaculture Society, Baton Rouge, Louisiana USA, pp 90–109

Wouters R, Van Horenbeeck T, Merchie G, Bridson P (2003b) Shrimp larval diets: dry diets made with fresh marine ingredients. Global Aquaculture Advocate 5(6):72–74

Wouters R, Van Horenbeeck T, Merchie G (2003c) Nutritional flakes for shrimp postlarvae. Asian Aquaculture Magazine, January/February 2003, pp 25–27

Wouters R, Cobo ML, Dhont J, Wille M (2009) Developments in feed formulations, feeding practices and culture techniques for marine shrimp larvae. In: Browdy CJ, Jory DE (eds) The rising tide: proceedings of the special session on sustainable shrimp farming, Aquaculture 2009. WAS, Baton Rouge, pp 79–91

Chapter 6
Aquaculture Production Systems and Environmental Interactions

Dror Angel, Alfred Jokumsen, and Giuseppe Lembo

Introduction

The strategic guidelines for the sustainable development of European aquaculture (COM(2013) 229 final) indicate that aquaculture can contribute to the overall objective of filling the gap between European Union (EU) consumption and production of seafood in a way that is environmentally, socially and economically sustainable. To fulfil this aim, each member state is encouraged to indicate in the multiannual national strategic plan its own aquaculture growth objective (volume and value) for the period covered by the plan. Also the current Common Fisheries Policy (CFP) aims to encourage aquaculture development in all European states, based on the strategic guidelines and the national multiannual strategic plans, identifying common objectives and, where possible, indicators to measure success.

One of the key factors to successful aquaculture is the availability and maintenance of high-quality marine and fresh waters. The Marine Strategy Framework Directive (MSFD) stipulates that EU aquaculture must comply with high health, consumer protection and environmental sustainability standards. These have cost implications for producers but can be turned into a competitive advantage if the products are

D. Angel (✉)
Department of Marine Biology and Maritime Civilizations,
Charney School of Marine Science, University of Haifa, Haifa, Israel
e-mail: adror@research.haifa.ac.il

A. Jokumsen
National Institute of Aquatic Resources, (DTU Aqua), Technical University of Denmark,
Nordsøen Forskerpark, Hirtshals, Denmark
e-mail: ajo@aqua.dtu.dk

G. Lembo
Stazione Sperimentale per lo Studio delle Risorse del Mare,
COISPA Tecnologia & Ricerca, Bari, Italy
e-mail: lembo@coispa.it

G. Lembo, E. Mente (eds.), *Organic Aquaculture*,
https://doi.org/10.1007/978-3-030-05603-2_6

marketed under an eco-label (and priced accordingly) and can also contribute to improved social acceptance of aquaculture.

The Blue Growth Initiative is a long-term programme to achieve the goals of the Europe 2020 strategy for smart, sustainable and inclusive growth to the benefit of all segments of society and the environment. Developing sectors that have a high potential for sustainable jobs and growth, such as aquaculture, is one of the three components on which the Blue Growth Strategy relies. The widespread acceptance of the ocean economy and Blue Growth ideas have moved them up in the international policy agenda.

At the Rio+20 Summit, for the first time, the conservation and sustainable use of the oceans were addressed along with the world's other most pressing sustainability challenges. One of the Sustainable Development Goals drafted by the Summit was to protect the oceans and to use marine resources sustainably (SDG 14).

The EU is committed to adopt this approach and to improve the way that oceans are managed. This will ensure that marine resources are used sustainably, for healthy marine ecosystems and a strong blue economy (SWD (2017) 128 final).

Indeed, aquaculture, which has a lower carbon footprint when compared to cattle, pork and even poultry husbandries (Sonesson et al. 2009), has the potential to effectively address the sustainability challenges and to increase social acceptance of aquaculture production systems and products. Nonetheless, EU organic aquaculture needs to emphasize the choice of technological solutions and science-based production systems and protocols in order to consolidate those sustainable aquaculture systems that comply with circular economy principles and maintain healthy aquatic ecosystems.

The most widely used production systems in aquaculture are "flow through" (FT) where water quality is maintained in the cultivation systems by continuous flushing/flow through of new water. These systems include (i) floating or submersible net cages or pens that may be placed in the sea, rivers or lakes; (ii) ponds that receive water from an adjacent river, or tidal areas, or coastal lagoons; (iii) raceways that receive fresh water or seawater; and (iv) tanks of various shapes and sizes, made mostly from fibreglass and plastic and used in hatcheries and nurseries where water quality is maintained by a continuous supply of new water.

In tanks or ponds, the water exchange rate is usually set to maintain the optimal water quality to enable the reared species to grow and to remain healthy. The water quality variables that determine the flushing/water flow rate vary as a function of the requirements of the species and may include water temperature, pH, alkalinity, dissolved oxygen, turbidity, ammonia, etc.

The farmed species in land-based FT systems generally benefit from an ample supply of dissolved oxygen and removal of wastes. However, the quality and the quantity of the available in-flow water may vary due to seasonal variations, weather conditions, etc. In many areas where space is a limiting factor, aquaculture has transitioned to marine aquaculture or mariculture where the space and supply of good quality water are less limiting. Sea cages or net pens exploit the ecosystem services provided by the sea where they are moored, and the farmed species benefit from the marine environment, which provides (dissolved) oxygen and disperses/transforms the wastes in the surrounding water.

The primary concern of net-cage farmers is the selection of the optimal site, which should be characterized by minimal environmental and other risks (ships, predators, rough seas, harmful algal blooms, pollution), optimal water exchange (currents) to provide good water quality in the cages and avoidance of self-pollution, i.e. to avoid re-entry of the water that was flushed out of the cage. Optimal aquaculture site selection was the focus of the EU AquaSpace project (http://www.aquaspace-h2020.eu/), involving modelling technology as a major approach used to select the best sites for this purpose.

Recirculation technology offers the potential both for improved farming operations and for reduced environmental impact. Recirculating aquaculture systems (RAS) are therefore considered to be a promising technology for sustainable fish farming, making it possible to increase production volumes and simultaneously reduce the environmental impact from aquaculture (Dalsgaard et al. 2012).

One of the major issues that remains to be solved is the maintenance and management of water quality in the cultivation systems. In RAS, as in all farming operations, stable and controllable water quality is of paramount importance for fish performance and welfare. A few studies (e.g. Pedersen et al. 2012) have investigated the effect of feed loading on water quality, and recent studies (e.g. Bentzon-Tilia et al. 2016; Dalsgaard et al. 2017; Rojas-Tirado et al. 2017) consider microbial factors that affect water quality in RAS.

RAS are generally intensive and often hyperintensive systems, i.e. they rely on substantial input of external energy, high stocking densities and disconnection of the aquaculture production from the natural aquatic environment. As such, they are not in line with the principles of organic farming but will be described below because they are very popular aquaculture production systems.

Integrated multi-trophic aquaculture (IMTA) is a relatively recent innovation in Europe, and one of its objectives is to employ extractive species, such as macroalgae and shellfish to take up and absorb particulate and dissolved matter from the aquaculture effluents. In this way, waste products (nutrients) from one species (e.g. finfish) enable the cultivation and harvest of a second and/or third species while reducing the load of nutrients and particles in the surrounding water and on the seafloor. This is often described as a "win-win" system, which enables farmers to comply with regulations regarding waste effluents. It also enables the producers to basically "reuse" the expensive feed offered to the "fed" species, i.e. there are environmental, regulatory and economic incentives in employing IMTA.

Flow-Through Systems

Flow-through systems are the most common systems currently used for intensive rearing of a wide variety of farmed species. Traditional land-based flow-through farms use water that is taken from an adjacent river or lake, by gravity or by pumps.

The water may have a wide variety of residence times in the system, and various water treatment units are subsequently used for mechanical (micro-sieves, settling ponds) and biological filtration of the farm water before discharging this into the receiving waters, downstream.

Originally, ponds were dug directly into the soil of river valleys close to the river or stream banks, but nowadays most farms have replaced earthen ponds with concrete tanks or other waterproof materials (e.g. in raceways).

The construction of fish ponds, tanks or raceways is designed to provide optimal water flow rates and physiochemical parameters to safeguard animal health and welfare and provide animals with sufficient space for their needs. Organic containment systems, where fresh water species are reared, require that the bottom type be as close as possible to natural conditions and, in the case of, e.g. carp and some mullet species, the bottom should be natural soil.

Land-based FT systems have undergone significant technological improvements in terms of reducing fresh water intake and reducing the environmental impact from the released effluents. Farms can therefore be classified, as is the case in Denmark, into three different types, according to the level of water consumption and wastewater treatment.

Type 1 is an extensive farm with mechanical water treatment and relatively low stocking density. It includes a minimal reuse of water (maximum use of 40 m^3 of water per kg of fish produced) and efficient internal (natural) assimilation of nutrients. Water treatment takes place partly by natural processes and partly via sludge cones, micro-sieves or contact filters, plant lagoons and sludge basins (Jokumsen and Svendsen 2010). This type of farm is clearly in line with the principles of organic farming.

Type 2 is a more intensive farm with both mechanical and biological water treatment, lower water consumption and increased reuse of water compared to farm type 1. In addition to mineralization of nutrients, water treatment occurs via sludge cones, micro-sieves, biofilters and sludge basins, but no plant lagoons are required. This type of farm might still be in line with the principles of organic farming provided that a series of obligations are met (e.g. the raised species spend the majority of their lives in outdoor facilities, at stocking density on average lower than in intensive aquaculture systems).

Farm type 3 is a hybrid – not really a FT farm as it is characterized by significantly lower water consumption (<3.6 m^3 water per kg of fish produced) than in traditional flow-through systems. This type of farm employs a high degree of recirculation technology, i.e. water reuse and wastewater treatment, which makes the system less reliant on the surrounding natural environment (Jokumsen and Svendsen 2010). As a result, reliance on local or adjacent water sources and on natural "purification" processes is minimal, and the impact of such farms on natural fauna/flora is therefore reduced.

Recirculating Aquaculture Systems (RAS)

Intensive recirculating aquaculture systems use a wide range of wastewater treatment devices (Martins et al. 2010; Dalsgaard et al. 2012). A sketch of a RAS is given in Fig. 6.1.

The water supply for an intensive RAS fresh water farm is typically groundwater; in the case of marine farms, the water used is pumped directly from the sea by means of submersible pumps or may be artificial seawater. RAS can be equipped with different technologies depending on specific contexts and requirements. A possible operating mode is described below. The production water from the fish tanks (1) passes through a mechanical filter (2), i.e. a micro-sieve (mesh size generally about 60 μm). The micro-sieve separates particulate matter, which is flushed as sludge to a sludge storage tank until it can be used as agricultural fertilizer or for production of biogas. From the micro-sieve (2), the water is pumped (3) to the biofilters (4), where the dissolved fractions, especially ammonia (NH_4^+), are converted into nitrate (NO_3^-). In a separate biofilter (5) with anoxic conditions (a denitrification filter), the NO_3^- is converted into N_2 gas in the presence of easily degradable organic matter (Van Rijn et al. 2006). The recirculated water passes on to a trickling filter (6) for degassing (N_2, CO_2) and aeration (oxygenation) before it enters (7–8) the fish tanks. Before entering the fish tanks, the water passes through an UV radiation device (9) to kill microorganisms, especially bacteria, in order to minimize disease. Some of the aerated water from the trickling filter is pumped through an oxygen cone (7–10) for additional oxygen enrichment before it enters (11) the fish tanks. In addition, pure oxygen may be added to each tank/section (Chen et al. 2006, Pedersen et al. 2012; Van Rijn 2013), and the temperature can be adjusted using devices for heating or cooling the water. The amount of new water added to the RAS corresponds to the amount required to flush the micro-sieves (2) and the biofilters (4), to compensate for evaporation and to keep the temperature and salinity at an

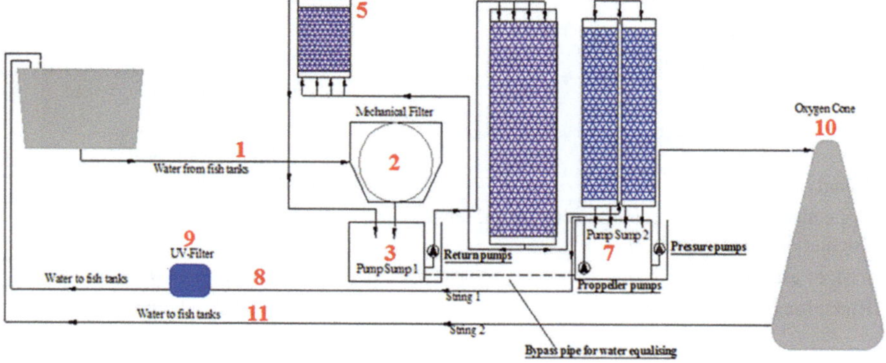

Fig. 6.1 Sketch of a recirculating aquaculture system (RAS). The numbers in the figure are referred to (in brackets) in the text. (Source: Billund Aquakulturservice ApS, Denmark)

appropriate level. The water consumption in RAS is more than 100 times less, i.e. less than 500 l/kg feed fed to the fish than in traditional flow-through systems (Jokumsen and Svendsen 2010).

Feed is the main source of nutrients and organic matter in RAS. Those fractions of the feed that are not utilized by the fish end up as a waste (dissolved or particulate) in the system and need to be processed by the recirculation technology and/or are discharged. The composition of the feed, the feeding management and the utilization of the feed by the fish and the concomitant wastewater treatment are therefore of utmost importance for RAS water quality and for the final discharge from these systems.

RAS rely on the input of external energy for circulating water within the system, water treatment and aeration of the water, as well as that required in the building that houses the RAS. The advanced technologies, management, comprehensive surveillance systems, working processes and hygienic procedures in a RAS farm require well-educated and well-trained personnel with the competence required to achieve optimal productivity. The high degree of recirculation makes it critical to continuously monitor and control the water quality within narrow limits, and the extensive use of alarm systems is necessary to monitor several key water quality parameters (Jokumsen and Svendsen 2010).

In Table 6.1, a comparison has been set up between a traditional flow-through system in organic farming and a recirculating aquaculture system (RAS).

Table 6.1 Advantages and disadvantages of a flow-through organic system and a recirculated aquaculture system

Flow-through organic system	RAS
Advantages	**Advantages**
Production in common with nature	Low water consumption
Favours biological diversity and animal welfare	Recycling of water
Natural temperature and light conditions	Stable farming conditions/water quality
Lower stocking density	Control of water temperature
Behavioural needs can be met	No environmental impact
Renewable energy use, e.g. for aerators	Prevents ingress of pathogens
Environmentally sustainable	Prevents escapes
–	Recycling/collection of waste nutrients (fertilizer)
–	Easy to disinfect/clean
Disadvantages	**Disadvantages**
Dependent on external conditions (weather, temperature fluctuations, water quality)	Energy consumption
Risk of escape	Use of costly pure oxygen
Risk of ingress of pathogens	Risks related to high stocking densities
–	In case of disease, risk of boosting prevalence

Source: EGTOP Aquaculture Report (Part B)

As shown in Table 6.1, RAS have several environmental advantages but require a significant input of external energy, relying on high stocking densities (for economic reasons), advanced wastewater treatment devices, use of UV radiation and use of pure oxygen. All the above, in addition to the disconnection of the aquaculture production from the external natural aquatic environment, makes RAS not in line with the provisions of the majority of the organic aquaculture standards and the EU regulations on organic aquaculture.

However, due to the limitations of water resources, an alternative strategy might be the so-called reuse of water, which, to some extent, combines the advantages of both flow-through systems and RAS, without compromising organic principles. Reuse of water means a kind of *extensive* recirculation in *outdoor* systems (Colt 2006). Instead of being discharged, the water is pumped back to the inlet and reused in the fish ponds, tanks or raceways after passing wastewater treatment devices such as natural filter beds, settlement ponds and mechanical or biological filters to collect waste nutrients and/or using seaweeds and/or bivalves and algae, which contribute to improving the quality of the effluent. The types and capacity of wastewater treatment devices depend on the specific conditions of the farm, the related production capacity/intensity and the fulfilment of water quality requirements.

In order to comply with the above-mentioned organic standards and regulations, the proper oxygen saturation in the aquatic environment should be achieved by using mechanical aerators. This requires careful planning of such farming systems so that a well-balanced equilibrium is established between the stocking density, the efficiency of the wastewater-nutrient removal and the amount of water reused for the proper operation of the organic farm.

Extensive Pond Systems

The most traditional extensive aquaculture system is based on fish ponds, i.e. a man-made aquatic ecosystem that is an integral part of the landscape (Kořínek et al. 1987; Adámek et al. 2012). Fish ponds are an integral part of the rural economy in some regions and are an almost natural part of the landscape because of their large number and often large size. Ponds were often established in low-lying wetlands or in areas with soil conditions that were too poor to support conventional agriculture.

Fish ponds initially relied on the natural pond ecosystem to provide the stocked fish with feed, and production levels were low. Modern fish ponds have greatly increased production levels through a variety of management practices (e.g. fertilization and supplementary feeding). These practices, in addition to the influence of agriculture and other human activities, have led to an eutrophic state of many ponds in Central Europe (Pechar 2000; Potužák et al. 2007; Všetičková et al. 2012). As a result, fish pond farmers must face the challenge of not only maintaining good health and nutritional status of fish but also the challenge of maintaining pond biodiversity as well as good water quality (Dulic et al. 2010; Máchová et al. 2010; Filbrun and Culver 2013).

Carp aquaculture is very popular in Central Europe. Carp ponds are carefully designed to match the needs of the carp at different life stages. Nursery ponds may be as large as 1 ha, with a water depth of only 0.5 m, and are used to rear carp up to the age of 1 year. Grow-out ponds are larger, up to 10 ha, and are used for fish from age 1–2 years. Carps are generally marketed when they reach 1.5–3 kg (3–4 years old), and marketing ponds are even larger (e.g. 50–100 ha).

In order to address some of the ecological problems that develop in ponds with considerable stocks of fish and an abundance of food and waste products, additional species of finfish may be stocked, and the monoculture carp pond becomes a polyculture pond. In addition to the common carp (*Cyprinus carpio*), often stocked at 50–90% of the species mix, bighead carp *Hypophthalmichthys nobilis*, grass carp *Ctenopharyngodon idella* and silver carp *Hypophthalmichthys molitrix* are stocked at 10–30% of the total abundances. In some cases, predators (e.g. pike *Esox lucius*, pikeperch *Stizostedion lucioperca*, European catfish *Silurus glanis*) and other species such as tench (*Tinca tinca*) may be added as a minor species so that all the components in the pond food web are efficiently exploited.

In addition to the better utilization of land resources, fish ponds may contribute to the management of water resources. Fish ponds are suitable not only for the production of fish but also to accumulate water, which can be used for irrigation during dry periods. Moreover, ponds support nature in the surrounding biotopes. Fish pond sediments accumulate organic matter and nutrients, and these can serve as an agriculture fertilizer that may be harvested when ponds are dried between growing cycles. In this manner fish farming can support horticulture or the production of other terrestrial plants.

In addition to the above, polyculture ponds can be integrated with intensive rearing of other fishes in adjacent facilities. In this case, effluents from flow-through raceways of intensive systems, e.g. for trout cultivation, may be discharged into polyculture ponds where the fish faeces and nutrients may increase the natural fish food production in the same way as manure application. If the ratio of supply from the intensive unit to the polyculture pond area is designed properly, a high level of water purification can be achieved.

Cage Systems

By far the largest proportion of European aquaculture biomass production takes place in cage and raft systems in the sea. Benefits of cage rearing are relatively low investment costs, low energy costs, efficient utilization of environmental resources (such as space and natural flushing of the cages) and a low carbon footprint compared to other protein production systems.

Cage farms were traditionally situated in safe (in terms of exposure to rough seas) and easily accessible locations, generally near shore. Whereas these locations were great for the farmers, the selection of shallow sites with limited flushing often resulted in organic overloading of the sediments under the farms and substantial negative effects (e.g. hypoxia, anoxia, sulphide and methane degassing, etc.) on seafloor biogeochemistry and on biodiversity. In addition to geochemical problems, parasitic and pathogenic outbreaks in fish farms may occur, challenging the farmers and potentially threatening wildlife outside the net pens (e.g. Thorstad and Finstad 2018). Moreover, intensive cage farming of salmon has led to significant increases in the density and occurrence of two sea lice species *Lepeophtheirus salmonis* and *Caligus elongatus*, which are ectoparasites infecting salmonids in seawater and causing severe mortalities in sea cages (Nilsen et al. 2017; Jackson et al. 2017).

Numerous studies have documented the negative environmental impacts associated with sea-cage aquaculture (e.g. Cole 2002; Nash et al. 2005; Huntington et al. 2006; International Union for Conservation of Nature 2007).

Further analyses of the impacts associated with fish farms, including meta-analysis of 30 peer-reviewed articles (Sarà 2007), found that dissolved oxygen was generally not affected by aquaculture operations. It has been shown that improvements in feed formulations, feed delivery (husbandry) and feeding efficiency are key factors for reducing nutrient loading and impacts to the quality of water close to sea-cage farms. Moreover, siting fish farms in areas of high hydrodynamics and strong water currents, with depths at least twice that of the net pen, is an additional factor that may contribute to a good water quality (Beveridge 2004; Belle and Nash 2008).

The response of marine benthic communities to enrichment is a function of site-specific environmental factors (e.g. Pearson and Rosenberg 1978), which should be considered prior to the selection of an aquaculture site. Indeed, the accumulation of fish and feed waste below cages and the associated geochemical changes can induce changes to the micro- and macrofauna that live in the sediments below sea-cage farms.

Sea cages should be designed and maintained in accordance with ambient conditions and operated so that there is a minimal risk of cage failure and fish escapes. The nets should routinely be inspected, and all necessary repairs or adjustments to anchors, ropes and nets should be carried out without delay. Access, proximity to hatcheries or fishing harbour, security, economic, social and market considerations should also be taken into consideration. In order to keep birds, seals and other predators away from cages and other aquaculture gear, non-lethal antipredator measures should be implemented, such as overhead netting or screens to exclude sea birds from cage areas (Nash 2001; Huntington et al. 2006; Halwart et al. 2007).

Antifouling measures are essential to limit the growth of marine organisms which attach to aquaculture cages, ropes and structures. Heavy and persistent biofouling impedes water flow-through cages, increases biological oxygen demand in the cages, causes net drag and can shorten the effective lifespan of nets and ropes. An alternative to chemical antifouling methods is to manage biofouling by letting the nets air-dry in order to kill the biofouling organisms. The nets can also be manually cleaned on land using large-scale net washers, or high-pressure spray, or scraping to remove encrusted organisms. The employment of grazers, cleaner fish, sea urchins and mechanical robots has also been proposed as some alternative to chemical treatments inside fish cages to remove attached organisms (Willemsen 2005; International Union for Conservation of Nature 2007).

Fallowing is a practice that has been recommended to prevent cumulative damage to the benthic environment and for reducing risks related to fish pathogens and parasites. Fallowing refers to the practice of relocating marine fish cages or discontinuing the production, for a certain time, after each production cycle, to allow the sediment below the cages to undergo natural recovery, both geochemically and ecologically, from the impacts of organic matter and nutrient loading.

Although sea cages are often associated with negative effects, these may also provide benefits to the local and regional environment by providing shelter and foraging habitat for wild fish (International Union for Conservation of Nature 2007; Grigorakis and Rigos 2011), serving as hatcheries and nurseries (Ozgul and Angel 2013). A further significant role that sea-cage farms may play in marine fisheries conservation, e.g. buffer against fishery mismanagement (such as overfishing), is that the farms may be designated as marine protected areas or no-take zones (Dempster et al. 2006) providing wild fish with safe havens from fishing operations.

Integrated Multi-trophic Aquaculture (IMTA)

Integrated aquaculture has been practised for centuries in fresh water systems, particularly in Asia, as an efficient means to rear several products simultaneously. There has been some confusion between polyculture (several species of finfish generally cultivated in ponds) and integrated multi-trophic aquaculture (IMTA – see definition, below), but this concept is not new. Although the potential benefits of such systems are now well known, IMTA is currently only practised to a limited extent in the EU.

IMTA may be applied to land-based (Gordin et al. 1981), coastal (Fang et al. 2016) and offshore (Buck et al. 2018) aquaculture systems in both marine and fresh water environments. IMTA offers a balanced ecosystem management approach that could benefit the farmers, consumers and environment.

IMTA represents a solution for a greater environmental stewardship and potentially better economic performance of aquaculture. It is a different way of thinking about aquatic food production that is based on the concept of circular economy. Instead of growing only one species (monoculture) and focusing primarily on the needs of that species, IMTA mimics a natural ecosystem by combining the farming of multiple, complementary species from different trophic levels in the food web. For example, fish, invertebrates (e.g. mussels, oysters, etc.) and seaweeds might be farmed in the same area, in a way that allows the uneaten feed, wastes, nutrients and by-products of one species to be recaptured and converted into feed and energy for the growth of the other species.

The natural ability of these species to recycle the nutrients provides environmental service improving the ecological performance of the aquaculture system. Moreover, in addition to their recycling abilities, the farmed species provide extra economic benefits to the farmers in terms of added value as marketable products. The added value is both the result of added biomass and diversity, and if the products are labelled as ecologically friendly, this may increase their retail market value and provide farmers with even greater profits.

The selected species and system design should be optimized in order to maximize recycling abilities, in both space and timing of culture. Indeed, larger organic particles, such as uneaten fish feed and faeces, may be consumed by deposit feeders, such as sea cucumbers and sea urchins placed below the sea cages. The fine suspended particles can be extracted from the water column by filter-feeding species, like mussels and oysters, on long lines situated around the farm. In order to efficiently capture dissolved nutrients (nitrogen, phosphorus) released from sea cages, seaweeds should be placed in the water column at a distance downstream from the farm. All of these "filtration" functions may be modelled to optimize the uptake of farm effluents for maximal growth and nutrient delivery.

IMTA also offers the potential to reduce the use of drugs involved in disease prevention and management. It has been shown, for example, that shellfish (e.g. mussels) may effectively reduce viral, bacterial and/or parasitic diseases in the cultured fish, due to their filtration activities. While this aspect is promising, there is also the risk that some IMTA species could act as intermediate hosts for disease agents and thus increase the risk to fish health.

In many monoculture farming systems, the fed aquaculture species (e.g. finfish) and the extractive aquaculture species (bivalves, herbivorous fishes and aquatic plants) are independently farmed in different locations. According to Barrington et al. (2009), IMTA is the practice which combines, in the appropriate proportions, the cultivation of fed aquaculture species (e.g. finfish/shrimp) with organic extractive aquaculture species (e.g. shellfish/herbivorous fish) and inorganic extractive aquaculture species (e.g. seaweed) to create balanced systems for environmental sustainability (bio-mitigation), economic stability (product diversification and risk reduction) and social acceptability (better management practices).

The multi-trophic subsystems are integrated in IMTA that refers to the more intensive cultivation of the different species in proximity of each other, linked by nutrient and energy transfer through the water (Fig. 6.2).

A number of issues arise in attempting to satisfy basic IMTA requirements in a commercial system. Environmental sustainability is the major consideration in IMTA; therefore the criteria guiding species selection are the imitation of natural ecosystems. Combinations of cocultured species will have to be carefully selected/balanced according to a number of conditions and criteria as described in Angel and Freeman (2009):

Complementary roles with other species in the system. Species should be able to feed on the other species' wastes in order to improve the quality of the water and grow efficiently. Not all species can be efficiently grown together.

Adaptability in relation to the habitat. Native species that are well adapted within their normal geographic range should be selected. This will help to prevent the risk of invasive species impacting the local environment and economic activities.

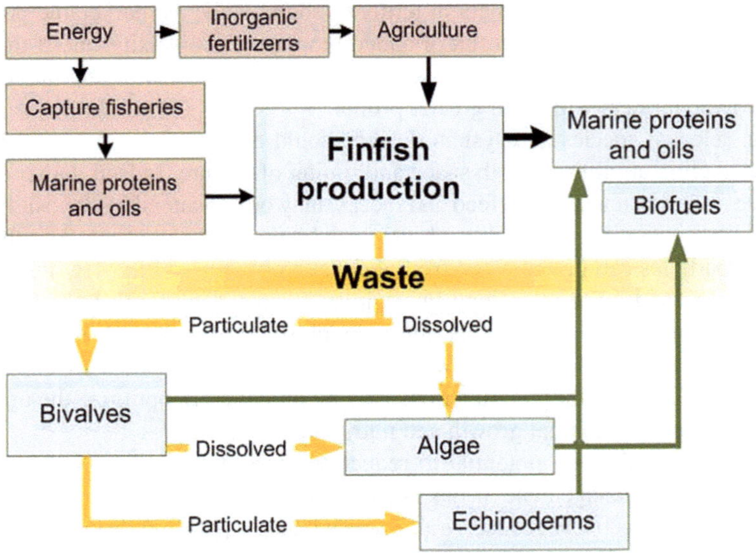

Fig. 6.2 Integrated multi-trophic aquaculture (IMTA) flow chart. (Source: IDREEM EU project – final report)

Culture technologies and site environmental conditions. The background levels of particulate organic matter and dissolved inorganic nutrients should be considered, as well as the size range of particles, when selecting a farm site.

Ability to provide both efficient and continuous bio-mitigation. Species that are capable of growing to a significant biomass should be selected. This feature is important if the organisms are to act as a biofilter that captures the excess of nutrients emitted from the farm.

Market demand for the species and pricing as raw material or for their derived products. Farmers should select species that have an established or perceived market value and be able to sell the extractive species in order to increase their economic outcome. Therefore, they should explore the markets in advance before investing too heavily.

Commercialization potential. Farmers should select species, for which regulators and policymakers will facilitate the development of new markets, and not impose new regulatory impediments to commercialization.

The beneficial environmental effects of sea-cage mariculture of finfish, integrated with filter-feeding bivalves, such as mussels and oysters, and seaweed are well documented (Angel and Freeman 2009):

Effluent bio-mitigation. Mitigation of effluents through the use of biofilters, which are suited to the ecological niche of the aquaculture site, can solve a number of environmental challenges posed by monoculture aquaculture.

Increased profits through diversification. Increased overall economic value of an operation from the commercial by-products that are cultivated and sold. To make environmentally friendly aquaculture competitive, it is necessary to raise its revenues. By exploiting the extractive capacities of cocultured lower trophic level taxa, the farm can obtain added value products that can outweigh the added costs involved in constructing and operating an IMTA farm. The waste nutrients in integrated aquaculture are not considered a burden but rather a resource, for the auxiliary culture of the extractive species.

Improving local economy. Economic growth may be realized through employment (both direct and indirect) in production, processing and distribution.

Form of "natural" crop insurance. Product diversification may offer financial protection and decrease economic risks when price fluctuations occur or if one of the crops is lost to disease or inclement weather.

Disease control. Prevention or reduction of disease among farmed fish can be provided by certain seaweeds due to their antibacterial activity against fish pathogenic bacteria.

Increased profits through obtaining premium prices. Potential for differentiation of the IMTA products through eco-labelling or organic certification programmes.

Nonetheless, there are also some challenges in the IMTA approach (Angel and Freeman 2009):

Higher investment. Integrated farming in open-sea settings requires a higher level of technological and engineering sophistication and upfront investment.

Difficulty in coordination. If practised by means of different operators (e.g. independent fish farmers and mussel farmers) working in concert, farm operation would require close collaboration and coordination of management and production activities.

Increased requirement of farming area. While aquaculture has the potential to decrease pressure on natural fisheries and IMTA has specific potential benefits for the enterprises and the environment, fish farming competes with other users over coastal and marine habitats. Stakeholder conflicts are common and range from concerns about pollution and impacts on wild fish populations to site allocation and local priorities. The challenges for expanding IMTA practice are therefore significant although it can offer a mitigation opportunity to those areas where mariculture has a poor public image and competes for space with other activities.

Difficulty in implementation without open water leasing policies. Few countries have national aquaculture plans or well-developed integrated management of coastal zones. This means that decisions on site selection, licencing and regulation are often ad hoc and highly subject to political pressures and local interests. Moreover, as congestion in the coastal zone increases, many mariculture sites are threatened by urban and industrial pollution and accidental damage.

All of these issues must be addressed when designing and planning IMTA operations so that sustainability of this concept truly encompasses the full spectrum of social, economic and environmental considerations.

References

Adámek Z, Linhart O, Kratochvíl M, Flajšhans M, Randák T, Policar T, Masojídek J, Kozák P (2012) Aquaculture the Czech Republic in 2012: modern European prosperous sector based on thousand-year history of pond culture. Aquac Eur 37(2):5–14

Angel D, Freeman S (2009) Integrated aquaculture (INTAQ) as a tool for an ecosystem approach in the Mediterranean Sea. In: Soto D (ed) Integrated mariculture: a global review. FAO Fisheries and Aquaculture Technical Paper, vol 529, pp 133–183

Barrington K, Chopin T Robinson S (2009) Integrated multi-trophic aquaculture (IMTA) in marine temperature waters. In: Soto D (ed) Integrated mariculture: a global review. FAO Fisheries and Aquaculture Technical Paper, vol 529, pp 7–46

Belle SM, Nash CE (2008) Better management practices for net-pen aquaculture. In: Tucker CS, Hargreaves J (eds) Environmental best management practices for aquaculture. Blackwell Publishing, Ames, pp 261–330

Bentzon-Tilia M, Sonnenschein EC, Gram L (2016) Monitoring and managing microbes in aquaculture – towards a sustainable industry. Microb Biotechnol 9(5):576–584. https://doi.org/10.1111/1751-7915.12392

Beveridge M (2004) Cage aquaculture. Blackwell Publishing, Oxford

Buck BH, Troell MF, Krause G, Angel DL, Grote B, Chopin T (2018) State of the art and challenges for offshore Integrated Multi-Trophic Aquaculture (IMTA). Front Mar Sci 5:165. https://doi.org/10.3389/fmars.2018.00165

Chen S, Ling J, Blancheton JP (2006) Nitrification kinetics of biofilm as affected by water quality factors. Aquac Eng 34:179–197

Cole R (2002) Impacts of marine farming on wild fish populations. Final Research Report for Ministry of Fisheries Research Project ENV2000/08 Objective One, National Institute of Water and Atmospheric Research, New Zealand. Available at: https://www.google.com/url?sa=t&source=web&rct=j&url=https://www.fisheries.govt.nz/dmsdocument/3690/send&ved=2ahUKEwjuprHrsJXhAhXNh3AKHU0fDPAQFjAAegQIBBAB&usg=AOvVaw2e1E730k6rNCZz6h5bChrz. Accessed 27 Sept 2012

Colt J (2006) Water quality requirements for reuse systems. Aquac Eng 34:143–156

Dalsgaard J, Lund I, Thorarinsdottir R, Drengstig A, Arvonen K, Pedersen PB (2012) Farming different species in RAS in Nordic countries: current status and future perspectives. Aquac Eng 53:2–13

Dalsgaard J, Pedersen L-F, Pedersen PB (2017) Workshop on recirculating aquaculture Systems (III). Aquacult Eng 78(Part A):1–84

Dempster T, Sanchez-Jerez P, Tuya F, Fernandez-Jover D, Bayle-Sempere J, Boyra A, Haroun R (2006) Coastal aquaculture and conservation can work together. Mar Ecol Prog Ser 314:309–310

Dulic Z, Subakov-Simic G, Ciric M, Relic R, Lakic N, Stankovic M, Markovic Z (2010) Water quality in semi-intensive carp production system using three different feeds. Bulg J Agric Sci 16(3):266–274

EGTOP (2014) Final report on aquaculture (Part B). https://ec.europa.eu/agriculture/organic/eu-policy/expert-advice/documents/final-reports_en

Fang J, Zhang J, Xiao T, Huang D, Liu S (2016) Integrated Multi-Trophic Aquaculture (IMTA) in Sanggou Bay, China. Aquac Environ Interact 8:201–205. https://doi.org/10.3354/aei00179

Filbrun JE, Culver DA (2013) Can reduced provision of manufactured feed improve fish production efficiency in ponds? N Am J Aquac 75(1):64–76

Gordin H, Motzkin F, Hughes-Games A, Porter C (1981) Seawater mariculture pond – an integrated system. Eur Aquac Soc Spec Publ 6:1–13

Grigorakis K, Rigos R (2011) Aquaculture effects on environmental and public welfare – the case of Mediterranean mariculture. Chemosphere 855:899–919

Halwart M, Soto D, Arthur JR (2007) Cage aquaculture: regional reviews and global overview. FAO Fisheries Technical Paper No. 498, FAO, Rome, Italy. Available at: http://www.fao.org/3/a1290e/a1290e00.html. Accessed 27 Sept 2012

Huntington TC, Roberts H, Cousins N, Pitta V, Marchesi N, Sanmamed A, Hunter-Rowe T, Fernandes TF, Tett P, McCue J, Brockie N (2006) Some aspects of the environmental impact of aquaculture in sensitive areas. Final report to the Directorate-General Fish and Maritime Affairs of the European Commission, Poseidon Aquatic Resource Management Ltd., U.K. Available at: http://www.ec.europa.eu/fisheries/documentation/studies/aquaculture_environment_2006_en.pdf. Accessed 27 Sept 2012

International Union for Conservation of Nature (2007) Guide for the sustainable development of Mediterranean aquaculture. Interaction between aquaculture and the environment. IUCN, Gland. Available at: https://cmsdata.iucn.org/downloads/acua_en_final.pdf. Accessed 27 Sept 2012

Jackson D, Moberg O, Stenevik Djupevag EM, Kane F, Hareide H (2017) The drivers of sea lice management policies and how best to integrate them into a risk management strategy: an ecosystem approach to sea lice management. J Fish Dis. 2017 00:1–7. https://doi.org/10.1111/jfd.12705

Jokumsen A, Svendsen L (2010) Farming of freshwater rainbow trout in Denmark. DTU Aqua Report no. 219–2010

Kořínek V, Fott J, Fuksa J, Lellák J, Pražáková M (1987) Carp ponds of central Europe. In: Michael RG (ed) Manager aquatic ecosystem, Ecosystems of the World 29. Elsevier, Amsterdam

Máchová J, Valentová O, Faina R, Svobodová Z, Kroupová H, Mráz J (2010) Water pollution by carp originating from different pond management systems. Bull VÚRH Vodňany 46(1):31–38

Martins CIM, Eding EH, Verdegem MCJ, Heinsbroek LTN, Schneider O, Blancheton J, Roque dÒrbcastel E, Verreth JAJ (2010) New developments in recirculating aquaculture systems in Europe: a perspective on environmental sustainability. Aquac Eng 43:83–93

Nash CE (2001) The net-pen salmon farming industry in the Pacific Northwest. U.S. Department of Commerce. NOAA Technical Memorandum NMFS-NWFSC-49. Available at: http://www.nwfsc.noaa.gov/publications/techmemos/tm49/tm49.htm. Accessed 27 Sept 2012

Nash CE, Burbridge PR, Volkman JK (2005) Guidelines for ecological risk assessment of marine fish aquaculture. U.S. Department of Commerce. NOAA Technical Memorandum NMFS-NWFSC-71. Available at: www.nwfsc.noaa.gov/assets/25/6450_01302006_155445_NashFAOFinalTM71.pdf. Accessed 27 Sept 2012

Nilsen A, Nielsen KV, Biering E, Bergheim A (2017) Effective protection against sea lice during the production of Atlantic salmon in floating enclosures. Aquaculture 466(1):41–50

Ozgul A, Angel DL (2013) Aggregations of wild fish at net-cage fish farms in the Gulf of Aqaba-Red Sea; implications for fisheries management and conservation. Aquac Environ Interact 4:135–145

Pearson TH, Rosenberg R (1978) Macrobenthic succession in relation to organic enrichment and pollution of the marine environment. Oceanogr Mar Biol Annu Rev 16:229–311

Pechar L (2000) Impacts of long-term changes in fishery management on the trophic level water quality in Czech fish ponds. Fish Manag Ecol 7(1–2):23–31

Pedersen L-F, Suhr KI, Dalsgaard AJT, Pedersen PB, Arvin E (2012) Effects of feed loading on nitrogen balances and fish performance in replicated recirculating aquaculture systems. Aquaculture 338-341:237–245

Potužák J, Hůda J, Pechar L (2007) Changes in fish production effectivity in eutrophic fishponds – impact of zooplankton structure. Aquac Int 15(3–4):201–210

Rojas-Tirado PA, Pedersen L-F, Pedersen PB (2017) Bacterial activity dynamics in the water phase during start-up of recirculating aquaculture systems. Aquac Eng. https://doi.org/10.1016/j.aquaeng

Sarà G (2007) Aquaculture effects on some physical and chemical properties of the water column: a meta-analysis. Chem Ecol 23:251–262

Sonesson U, Davis J, Ziegler F (2009) Food production and emissions of greenhouse gases. Se2009.eu, p 26

Thorstad EB, Finstad B (2018) Impacts of salmon lice emanating from salmon farms on wild Atlantic salmon and sea trout. NINA Rep 1449:1–22

Van Rijn J (2013) Waste treatment in recirculating aquaculture systems. Aquac Eng 53:49–56

Van Rijn J, Tal Y, Schreier HJ (2006) Denitrification in recirculating systems: Theory and applications. Aquac Eng 34:364–376

Všetičková L, Adámek Z, Rozkošný M, Sedláček P (2012) Effects of semi-intensive carp pond farming of discharged water quality. Acta Ichthyol Piscat 42(3):223–231

Willemsen PR (2005) Biofouling in European aquaculture: is there an easy solution? European Aquaculture Society Special Publication No. 35. Available at: http://www.crabproject.com/client/files/Paper_Willemsen.pdf. Accessed 2 Oct 2012

Chapter 7
Welfare Issues and Veterinary Treatments

Giuseppe Lembo, Pierluigi Carbonara, Andrea Fabris, Amedeo Manfrin, and Walter Zupa

Introduction

Among public and governments, there is increasing interest in the welfare of farmed fish. In addition, among farmers, there is growing awareness that good welfare equates to increased success of production activities (Lembo and Zupa, 2010). However, animal welfare is not easy to be defined. It is generally referred to the physical and mental state of the animal that is interacting with its environment and its associated variations (Chandroo et al. 2004).

The primary basis for the concept of "animal welfare" is the belief that animals are sentient being capable to experience good or bad feelings or emotional states (Dawkins 1990). Stress and stress-related responses should be considered as an adaptive condition of the organism that has the fundamental function of preserving the individual's life. In addition, it is increasingly clear that individuality in stress reactions has to be included in the concept of animal welfare. Such differences often take the form of suites of traits, or stress coping styles (SCS), where traits like sym-

G. Lembo (✉)
Stazione Sperimentale per lo Studio delle Risorse del Mare,
COISPA Tecnologia & Ricerca, Bari, Italy
e-mail: lembo@coispa.it

P. Carbonara · W. Zupa
COISPA, Bari, Italy
e-mail: carbonara@coispa.it; zupa@coispa.it

A. Fabris
Director of the Italian Association of Aquaculture farmers, Verona, Italy
e-mail: info@api-online.it

A. Manfrin
IZSVe, Legnaro, Italy
e-mail: manfrin@izsvenezie.it

G. Lembo, E. Mente (eds.), *Organic Aquaculture*,
https://doi.org/10.1007/978-3-030-05603-2_7

pathetic reactivity, aggression and the tendency to follow and develop routines show positive relationships.

Most animal welfare definitions can be categorized into "function-based" or "feeling-based".

"Function-based" definitions basically assume that welfare is correlated with biological functioning, including physiological stress responses (Duncan 2005). This definition implies that if an animal is in good health and has proper functioning of bodily systems, it is experiencing good welfare.

"Feeling-based" definitions assume that the animal is in a good welfare if "…is free of negative experiences, such as pain, fear and hunger and has access to positive experiences, such as social companionship…" (Huntingford and Kadri 2008, 2009). Indeed, welfare barely equals the current emotional state of the animal (Duncan and Dawkins 1983) and, in the longer term, it represents the balance between positive and negative subjective experiences (Martins et al. 2012).

According to the Reg. (CE) 834/2007, recital 17, organic stock farming should respect high animal welfare standards and meet animals' species-specific behavioural needs, while animal-health management should be based on disease prevention. In this respect, particular attention should be paid to housing conditions, husbandry practices and stocking densities. Moreover, the choice of breeds should take account of their capacity to adapt to local conditions. The implementing rules for livestock production and aquaculture production should at least ensure compliance with the provisions of the European Convention for the Protection of Animals kept for Farming Purposes (T-AP) and the subsequent recommendations adopted by its standing committee on 5 December 2005.

Moreover, according to Reg. (CE) 834/2007, Art. 15 1(b)(ii), husbandry practices, including feeding, design of installations, stocking densities and water quality shall ensure that the developmental, physiological and behavioural needs of animals are met.

The five welfare domains specified by Mellor and Stafford (2001) and reworked by FSBI into a form that is more appropriate for fish are currently considered an acceptable framework for evaluating suffering of farmed fish (FSBI 2002 – Fish welfare; Lembo and Zupa, 2010). However, determination of animal welfare requires the selection, collection and interpretation of different parameters and validated indicators. The aspects of the fish's condition that are often used in this context are its health status, its physiology and its behaviour. A set of simple, nonintrusive signs or danger signals that can be easily used as indicator, without needing access to laboratory apparatus, is provided in Table 7.1.

How well these signs work in any given case will depend on the species concerned, on circumstances and also on individual status.

Farmed fish are exposed to a range of industry practices that may act as chronic stressors which potentially compromise welfare. The effects of a wide range of aquaculture practices on the stress physiology of fish are well documented and have been reviewed by Conte (2004) and Pickering (1991). Some of these practices include frequent handling, transport, periods of food deprivation, deteriorating water quality, suboptimal stocking densities, fin-clips and environmental enrichment (Ashley 2007; Huntingford et al. 2006; Schram et al. 2010).

Table 7.1 Welfare indicators of fish welfare

Welfare indicators for fish		Interpretation
1	Changes in colour	Stress-induced changes in skin or eye colour (with a complex hormonal background) could be a sign of exposure to adverse events and/or social stress/subordinate status
2	Changes in ventilation rate	A high oxygen demand is reflected by rapid irrigation of the gills. The rate of opercular beats may be increased by stress and can be counted automatically or by eye
3	Changes in swimming and other behaviour patterns	Fish may respond to unfavourable conditions by adopting different speeds of swimming and by using different regions of a tank or cage. These include excessive activity or immobility, body positions that protect injured fins, escape attempts in confined conditions and chafing movements to dislodge ectoparasites
4	Reduced food intake	Loss of appetite is potentially a sign of impaired welfare
5	Slow growth/loss of condition	Fish change shape and/or slow growth or sustained reductions in growth may be indicative of chronic stress or a possible sign of trouble
6	Morphological abnormalities	The occurrence of morphological abnormalities can be used as an indicator of poor larval rearing conditions
7	Injury	Injury (e.g. dorsal fin injury, scales dislodged rather than lying flat) may be a direct consequence of an adverse event and a sign of poor welfare. In addition, because immune responses can be suppressed by cortisol, slow recovery from injury (or a high incidence of injury) may be a sign of generally poor conditions
8	Disease status	Increased incidence of disease in any population of fish may be a warning signal that there may be other underlying problems

Reworked from FSBI (2002)

Water Quality

Aquaculture production systems, such as ponds, raceways, sea cages, flow-through (FT) and recirculating systems (RAS), all differ one from each other for the characteristic of the water, in terms of quantity, quality, temperature, etc. Water is the medium in which farmed fish have to meet both their physiological and spatial needs. The waste derived from fish feed and its metabolic end products, such as uneaten feed, faeces and excretions and dissolved inorganic nutrients, can seriously impair water quality and can cause stress, reduced growth and increased incidence of diseases to the point of being lethal for fish themselves.

Water quality is often referred to chemical and physical parameters, such as concentration of dissolved oxygen, carbon dioxide, un-ionized ammonia-nitrogen, nitrite-nitrogen, pH, nitrate concentration alkalinity and temperature (Masser et al. 1999; Losordo et al. 1999; Conte 2004).

Oxygen concentration is surely the most important environmental parameter for all fish species, in both freshwater and marine habitats. A reduced availability of dissolved oxygen in the rearing environment, together with higher presence of un-ionized ammonia, can be ascribed to a high fish density and to the feeding practices, as well as to algal blooms and elevated temperatures (Ellis et al. 2002). This lack of oxygen can induce the typical metabolic adjustments caused by the hypoxic stress in order to maintain oxygen concentration in the critical organs and to reduce its consumption (Heath 1995). Low dissolved oxygen and high un-ionized ammonia levels can act as chronic stressors, elevating plasma cortisol levels (Pickering et al. 1991) and modifying physiological, as well as morphological condition in farmed sea bass (Saroglia et al. 2002).

The demand of oxygen increases with increasing temperature. Dissolved oxygen is the first water quality parameter that may limit production both in freshwater and marine aquaculture. Its availability is dependent on temperature and CO_2, but the usual recommendation for cold-water species is that they will have adequate oxygen concentration as long as the dissolved oxygen does not fall below 80% (Wedemeyer 1996). The threshold oxygen concentration for growth in rainbow trout has been shown to be about 75% saturation (Pedersen 1987). For Atlantic salmon, the optimal concentration is from 80% to 100%, but they can cope with 60%, although for shorter periods.

Dissolved oxygen concentration is surely among the most important environmental parameters also for European sea bass and above all for gilthead sea bream. Its concentration contributes also to modulate fish sensibility to other water quality parameters. It was demonstrated, for example, that juveniles of gilthead sea bream exhibit increased sensitivity to ammonia in case of oxygen saturation drops below 85% of saturation, while increased mortality occurs when the saturation is below 40% (Wajsbrot et al. 1991; EFSA 2008).

Temperature is a limiting factor for fish growth through the effects on feeding and metabolism, which can be also differently expressed according to the life stage. Acute temperature changes represent a realistic risk in aquaculture facilities where temperature may act as a stressor, particularly due to accentuated diurnal temperature cycles in shallow ponds or tanks or due to accidental temperature shocks during water turnover. Under such conditions, the dissolved oxygen in the intensive cultures becomes a further interacting limiting factor.

Temperature could also influence the typical management operation in aquaculture facilities, such as sedation and anaesthetization. Mylonas et al. (2005) demonstrated that lower temperature resulted in significantly longer anaesthesia induction and recovery time, presumably due to the positive relationship between temperature and opercular ventilation rates and metabolism. Evidences were reported by Barnabé (1991) who observed that sea bass juveniles cease growing at 11–15 °C and grow fast at 22–25 °C. It has also been shown that fin condition may be affected by metabolic activity under the control of ecological factors, such as temperature and O_2 concentration acting as limiting factors (Person-Le Ruyet and Le Bayon 2009). In sea bass, fins result more eroded at elevated temperature than in cold water, as fish are less active, especially when feeding: meal duration is shorter and daily feed intake is lower that, in turn, is responsible of lower growth rate

(Person-Le Ruyet et al. 2004). This requires trade-off solutions because lowering temperature leads to lower fin damages but also to lower growth rates. Adult sea bass can withstand temperatures ranging from 2 to 32 °C (Barnabé 1990), although Claireaux and Lagardère (1999) quantified the metabolic performances as dependent by temperature. Thus temperature range 20–25 °C corresponds to the thermal optimum for better growth performance of the species.

Optimal rearing temperature in salmon production ranges between 8 and 14 °C (Marine Harvest 2014). Hyperthermic conditions, especially in the first stages of the salmon life, may lead to spinal deformities. In a study by Ytteborg et al. (2010), they documented spinal deformities in fish that hatched at 10 °C and were exposed to 16 °C during first feeding, as opposed to fish that hatched at 6 °C and were exposed to 10 °C at first feeding.

Stocking Density

According to the "Recommendation concerning farmed fish", adopted on 5 December 2005 by the Standing Committee of the European Convention on the Protection of Animals kept for Farming Purposes, the husbandry environment of aquaculture animals shall be designed in such a way that the animals shall be kept in water of good quality, with sufficient oxygen levels, in accordance with their species-specific needs. More specifically, the Commission Regulation (EC) N° 889/2008 establishes that fish shall be kept in temperature and light conditions that respect the requirements of the species, having regard to the geographic location. Furthermore, in order to consider the effects of stocking density on the welfare of farmed fish, the condition of the fish (such as fin damage, other injuries, growth rate, behaviour expressed and overall health) and the water quality shall be monitored.

Actually, rearing density encompasses a complex web of interacting factors, such as water quality, social interactions, fish-to-fish interaction and fish-to-housing interaction that can have an effect on many aspects of welfare (Ashley 2007; Turnbull et al. 2008). Although stocking density is a parameter that can be easily documented and controlled, it is considered only an indirect indicator of fish welfare, while a combination of welfare indices (e.g. behavioural and water-quality monitoring) would be a better way to monitor fish welfare in aquaculture than monitoring just one index (i.e. stocking density). Therefore, the compliance with stocking density threshold values in combination with optimal water quality parameters, e.g. oxygen and carbon dioxide concentrations, would make the fish welfare conditions more reliable in the rearing environment.

It is very difficult to make generalizations about how rearing density affects welfare in all situations, because a great interspecific variability could be associated with the responses to this factor (Turnbull et al. 2008; Ellis et al. 2002; Conte 2004). Nevertheless, high rearing densities induce in many teleost the increase of the energetic expenditure for basal life functions that in turn could become detrimental for growth and immune-resistance and could also affect the social interaction between fish (Huntingford 2004; Martins et al. 2012).

There is a large number of biotic and abiotic factors that may influence the metabolic rate of fish. Among the other physiological (e.g. digestion, reproduction) and environmental factors (e.g. circadian or seasonal cycles), the most relevant are body mass, water temperature and oxygen availability. The aerobic scope is a meaningful parameter indicating the global amount of energy available for living in fish and may be considered as a sort of measure of individual physiological state and well-being (Carbonara et al. 2010; Korte et al. 2007; Zupa et al. 2015). Indeed, fish approaching the aerobic scope during their life may impair their growth and survival (Claireaux et al. 2000; Norin and Clark 2016). In Table 7.2 is reported a summary of the literature data on standard metabolic rate (SMR), active metabolic rate (AMR) and scope for activity (SFA) of sea bass, sea bream, rainbow trout and Atlantic salmon.

Table 7.2 Standard metabolic rate (SMR), active metabolic rate (AMR) and scope for activity (SFA) values for sea bass, sea bream, rainbow trout and Atlantic salmon

Specie	Temperature (°C)	Weight (g)	SMR (mg/kg/h)	AMR (mg/kg/h)	SFA (mg/kg/h)	References
Sea bass	20	200 ± 10	85.55 ± 7.05	509.09 ± 12.35	413 ± 8.83	Chatelier et al. (2006)
	19	147.16 ± 2.68	91.84	337.92	246.08	Claireaux et al. (2006)
	18	258 ± 51.1	115.99	598.81	482.81	Zupa et al. (2015)
	18	420 ± 41.6	162.56	826.8	664.24	Zupa et al. (2015)
Sea bream	18	100–199	189.23	664.36	475.13	Zupa et al. (pers. comm.)
	18	200–299	118.71	650.27	531.56	Zupa et al. (pers. comm.)
	18	300–399	117.33	597.25	479.92	Zupa et al. (pers. comm.)
	18	400–600	84.64	498.78	414.14	Zupa et al. (pers. comm.)
Rainbow trout	10±1	230–631	126.02	705.59	579.57	Zupa et al. (pers. comm.)
	14 ± 0.1	205 ± 11	144 ± 2.24	684.48 ± 16.96	540.48	McKenzie et al. (2007)
	492 ± 44		118.4 ± 2.7	595.7 ± 34.5	477.3	Skov et al. (2011)
Atlantic salmon	3	479 ± 18	44 ± 5		212±7	Hvas et al. (2017)
	23	413 ± 13	231 ± 5		421±12	Hvas et al. (2017)
	4	1.114 ± 208	36.7 ± 8.4	250.6 ± 40.2	213.9	Lucas (1994)
	10	1.114 ± 208	72.8 ± 11.9	423.6 ± 25.2	350.8	Lucas (1994)

However, in order to correctly interpret the metabolic scope values, rather than the magnitude of these values, it is more relevant to assess how often fish use completely its own amount of metabolic scope during its life.

From a physiological point of view, high-density condition increases red muscle activity leading to a rise of the global scope for activity (Lembo et al. 2007). At a density of 50 kg/m^3, sea bass muscle activity measured as electromyogram activity (EMG) was on average twofold higher than in fish reared at 10 kg/m^3 (Carbonara et al. 2013). Haematological parameters are indicators of fish oxygen demand to maintain the basal metabolism. Haemoconcentration is, indeed, reported as a strategy for increasing oxygen-carrying capacity of blood during periods of high energy demand (Houston 1990), such as a stress event or an important swimming activity. Haematocrit, haemoglobin and red blood cell count have generally the higher levels at the higher densities (Carbonara et al. 2013). Physiological responses to stress are driven by an increase of plasma cortisol levels. There is evidence that, within certain limits, the cortisol concentration increases proportionally with the stress levels, just before downregulation control and saturation of the cortisol receptors occur (Mommsen et al. 1999). Santos et al. (2010) showed that increased density levels reduce feed intake and growth and that feed intake reduction is partially compensated by a decrease in maintenance requirements for energy at the highest density. Another feature of intensively reared sea bass is reported by Roncarati et al. (2006), for which plasma triglycerides, total cholesterol and transaminases were found to be always significantly higher than in semi-intensively maintained fish.

Some authors observed that, under high rearing density, both adult and juvenile sea bass grow slowly than under low rearing densities (Saillant et al. 2003a, b, D'Orbcastel et al. 2010). Montero et al. (1999) described the effects of high rearing density on juvenile sea bream reporting, as a consequence of the stressful condition, an increase of haematocrit, haemoglobin and red blood cell concentration and the decrease of the alternative complement pathway (ACP), an important component of the immune system of fish. This effect, in salmonid species, has been reported to be a consequence of an elevation of plasma cortisol. Indeed, high plasma cortisol levels produce an immunosuppressive effect in fish, reducing circulating lymphocytes and increasing the susceptibility of fish to disease. Moreover the authors observed the decrease in hepato-somatic index and an altered liver fatty acid composition. These alterations reflected the effect of stocking density on lipid metabolism channelled to increase the energy demand.

Differences in growth and welfare have been also reported for fish farmed in sea cages at high densities, which has been related with increase of plasma cortisol. Variations on fish interactions towards the net pen are associated with both rearing density and the condition of the net. Increasing stocking density results also in an exponential increase of the escape rate from cages. Particularly, sea bream increases net inspection and biting in relation to the rearing density and, in situation of limited feeding, is more attracted by damages on the net structure, presenting higher escape rate (Glaropoulos et al. 2012; Papadakis et al. 2013).

It is worthwhile to highlight that even low densities can affect welfare and behaviour. Indeed, Batzina et al. (2014) showed that sea bream reared at 4.9 kg m^{-3}

exhibited aggressive behaviour and size distribution, indicating that such low density created a less favourable social environment than specimens reared at 9.7 kg m^{-3}. The author reported also that the use of blue liner in the tanks enhanced growth, suppressed aggression and reduced brain serotonergic activity, demonstrating that substrate and density effects are socially induced.

Oppedal et al. (2011) compared the effect of normal (5.6–14.5 kg m^{-3}) and high (15.7–32.1 kg m^{-3}) stocking densities on production and welfare parameters of Atlantic salmon. As a result, fish under normal density had better condition factor, better SGR and better feed intake throughout the experiment. There were no differences in fin damage or body lesions until the last sample point (more fin damage and lesions in fish under the higher stocking density). Fish under the high stocking density had more cataracts than fish under normal densities.

The overall picture arising from the studies performed to date investigating the effects of stocking density on different parameters suggests that both low and high densities are potentially detrimental to welfare. Furthermore, the results of these studies clearly illustrate the complex nature of the interaction between stocking density and fish welfare, with several environmental factors. As a consequence, it is a complex undertaking to model these multiple interacting and confounding influences of stocking density on measures of welfare (Turnbull et al. 2008), in an effort to gain an overall understanding.

Transport, Killing and Slaughtering

The transport of live fish involves the transfer of large numbers (or biomass) of fish in a small volume of water. Some important environmental parameters could severely change during long transport, such as water temperature, oxygen and CO_2 concentration (Delince et al. 1987). Handling and confined spaces could generate hyperactivity conditions that could result in lactate accumulation and affect blood oxygenation capacity. Aquaculture practices, such as handling, crowding and transport, stimulate a response to stress that occurs with plasma catecholamines and corticosteroids, as well as with changes in the characteristics related to metabolism, hydromineral balance and cardiovascular, respiratory and immune functions (Barton 2002). The catecholamine and corticosteroid productions are also responsible, at cellular level, for the expression of heat shock or stress proteins.

Stocking density has a large effect on social interactions between fish. This is the passive nonaggressive behavioural interactions, such as fin erosion, body injury, collision and abrasion with conspecifics and the physical tank environment, as well as aggressive behavioural interactions between conspecifics that can be detrimental to welfare (Ellis et al. 2002).

Iversen and Eliassen (2009) looked at the effects of sedation (isoeugenol 2.5 mg L^{-1}) on the primary (plasma cortisol), secondary (osmoregulation) and tertiary (mortality) stress responses of Atlantic salmon smolts during transport. As a result of the experiment, control fish had significantly higher plasma cortisol levels than sedated fish, for up to 6 h after transport. Cortisol levels were significantly higher than pre-transport levels in both groups, until 12 h after transport. Control fish had significantly higher lactate than sedated fish, up to 1 h after transport, and significantly higher plasma magnesium levels than sedated fish, up to 168 h after transport. Mortality was 11.3% for control fish and 2.5% for sedated fish and stopped 16 days after transport. Authors concluded that isoeugenol is a promising stress-reducing sedative for Atlantic salmon smolts and if used properly could improve animal welfare and survivability during and after common aquaculture-related incidents. Iversen et al. (2008) also demonstrated similar results to above when using clove oil (90–95% eugenol).

An optimal slaughter method should render fish unconscious until death, without avoidable excitement, pain or suffering prior to killing. Welfare evaluation at time of slaughter is difficult to measure because it requires a multidisciplinary approach examining various indicators such as brain functions, endocrine responses, behaviour and post-mortem tissue biochemical condition (Poli et al. 2005).

The biochemistry of the muscle post-mortem and the onset of rigour are influenced by the method used in preslaughter handling, stunning and killing of fish (EFSA 2009a; Lowe et al. 1993), which can compromise the organoleptic qualities and marketability of the final product.

Short starvation before slaughter (generally 1–3 days) is commonly performed to empty the gut and to reduce the probability of fish being contaminated with feed and faeces during the subsequent slaughter procedure. This practice also induces a reduction in ammonia excretion by the fish, which could reduce the water quality deterioration occurring during crowding and transport to slaughter facilities (EFSA 2009b), while prolonged starvation can lead to immune depression, which makes fish more susceptible to stress-mediated diseases during the preslaughter period (EFSA 2009b). In sea bass long starvation periods could induce a decrease of the intestinal microvilli length with change in permeability of intestinal mucosa to amino acids. This induces also loss of weight and condition and loss of intestinal fats and plasma protein, together with a precocious involution of gonad tissue, without any variation in the chemical composition of muscle (EFSA 2009b).

Fish welfare is strongly affected during preslaughter phase because vigorous movements are induced by crowding conditions. Nevertheless, during preslaughter handling operations, crowding and confinement represent unavoidable practices, required to rapidly remove fish from rearing units.

EFSA report (2009a) concluded that stunning either percussive or electrically is the most humane method and that preslaughter treatment as crowding and pumping may cause harm to the fish. When percussion is applied, it should be measured whether the air pressure, which drives the bolt, is sufficiently high to induce immediate loss of consciousness and sensibility (Van de Vis et al. 2014).

In sea bass and sea bream, field recognition for unconsciousness or death includes absence of breathing and opercular movements, eyes fixed (i.e. eye roll absent), absence of response to painful stimuli (pin-prick) and loss of balance (EFSA 2009a).

The onset and development of rigor mortis is widely used as indicator of pre-mortem stress and is influenced by many factors, such as species, age and size of the specimen and preslaughter procedures (Lowe et al. 1993; Nakayama et al. 1999). Bagni et al. (2007) reported that rigour starts earlier in crowded fish. Indeed, the post-mortem metabolism varies considerably between stressed and unstressed fish, where ATP is more or less depleted, respectively (Berg et al. 1997).

In sea bream immersed in ice slurry, the response to handling and breathing all stops after 15–20 min, whereas carbon dioxide-stunned fish appear dead after 5 min (Giuffrida et al. 2007). Body temperature decreases faster in liquid ice than in conventional ice. Hence, this method results to be effective and faster than conventional ice, fish may be stressed less, and the method is easily adaptable to the farms needing (Urbieta and Ginés 2000). Moreover, the fish slaughtered with liquid ice show better texture and freshness characteristics (Zampacavallo et al. 2015).

Ice slurry method, although commonly used for both sea bass and sea bream, is not considered to be welfare-friendly because it does not induce immediate brain dysfunction. Electrical stunning can induce immediate loss of consciousness and sensibility in fish. However, reported data show that fish cannot be killed by the use of electricity, as the fibrillation of the heart is not permanent (Van de Vis et al. 2014). This implies that electrical stunning should be followed by a killing method to avoid recovery of the stunned fish. Because stunning and killing are procedures that take some time, it is normally necessary to apply the electrical current for a certain duration of time, so as to allow subsequent killing before the fish have recovered.

A problem of electrical stunning, especially when fish are immersed in water during stunning, is that carcass damage might occur, such as muscle haemorrhages or a broken vertebral column. Roth et al. (2009) found that this problem could be overcome by exposing fish to the electricity after draining the water, so-called dry stunning. In this method, the fish are exposed to an electrical current via a series of rows of positive-plate electrodes and a conveyor belt acting as the negative electrode. Correct voltage, current or electrical field depends on species, orientation of the fish and also conductivity of the water, duration of exposure, electrical field strength and frequency. Recommended amperages to achieve an instantaneous stun in Atlantic salmon are in Robb and Roth (2003), Roth et al. (2003, 2004) and Lambooij et al. (2010). However, the electrical stunning/killing methods, even if they reduced the time to stun, still do not appear to satisfy all quality requirements for Mediterranean species, such as sea bass and sea bream, as revealed by the early rigor mortis onset/release and the shortage of shelf life (Zampacavallo et al. 2015). Further studies are needed in order to make electrical stunning a suitable method also for the Mediterranean species. Indeed, the EFSA Animal Health and Welfare panel recommended the urgent development of commercial stunning methods to induce immediate (or rapid) unconsciousness in sea bass and sea bream (EFSA 2009b).

Electrical stunning/killing methods show a positive effect on the quality, with a very low incidence of injuries, if applied on Atlantic salmon and trout. In the UK many freshwater trout are electrically stunned before being placed directly in ice where they die by asphyxia before recovery (Lines and Spence 2012). Percussive stunning is done by giving a fish a blow to the head with a wooden or plastic club or by using an instrument.

According to the "Recommendation concerning farmed fish" delivered by the Standing Committee of the European Convention on the Protection of Animals kept for Farming Purposes, if fish are ill or injured to such an extent that treatment is no longer feasible and transport would cause additional suffering, they must be killed on the spot and without delay by a person properly trained and experienced in the techniques of killing, except in an emergency when such a person is not immediately available. The choice of the emergency killing method to be used depends on the farming system, on the species, on the size and on the number of fish to be killed.

Veterinary Treatments

Antimicrobial resistance (AMR) is an increasing problem for humans and terrestrial animals, while at the moment the use of antibiotics in aquaculture has led to limited development of antibiotic-resistant bacteria. On the other hand, the accumulation of antibiotics in the environment, resulting in water and soil pollution, is a real problem that must be fought. Thus, vaccination is the most effective and environmentally friendly approach to prevent diseases in aquaculture to manage fish health. As only a few commercial vaccines are available, autogenous vaccines, produced with the specific pathogen isolated from diseased animal in a farm, are a cost-effective opportunity to prevent disease outbreaks ensuring aquatic animal health and welfare.

The use of plants for vaccine production also offers several advantages such as low cost, safety and easy scaling. To date a large number of plant-derived vaccines, antibodies and therapeutic proteins have been produced for human health, of which a few have been made commercially available. The use of plants for the development and production of recombinant vaccines offers several advantages. Indeed, plant-based systems are more economical, as plants can be grown on a larger scale than in other systems. Moreover, natural plant products present a viable alternative to antibiotics and other banned drugs being safer for the reared organism and humans, as well as the environment (Kolkovski, 2011).

Recently, increasing attention is being paid to the use of plant products for disease control in aquaculture as an alternative to chemical treatments. Plant products have been reported to stimulate appetite and promote weight gain and be stress resistance boosters, to act as immunostimulant and to have antibacterial and anti-parasitic (protozoans, monogeneans) properties in fish and shellfish aquaculture, due to active molecules such as alkaloids, terpenoids, saponins, flavonoids, phenolics, polysaccharides and proteoglycans.

The use of medicinal plants in aquaculture has attracted a lot of attention globally and has become a subject of active scientific investigations (Bulfon et al. 2014). The most investigated herbs are those widely used in folk medicine in China, India, Thailand and Korea, such as *Achyranthes aspera*, *Angelica sinensis*, *Astragalus membranaceus*, *Azadirachta indica*, *Cynodon dactylon*, *Echinacea purpurea*,

Massa medicata, *Punica granatum*, *Solanum nigrum*, *Withania somnifera* and *Zataria multiflora*.

Other plants are used all over the world for both curative and culinary purposes, such as garlic, green tea, cinnamon, turmeric, lupine, mango, peppermint, nutmeg, basil, oregano, rhubarb, rosemary and ginger.

The herbal remedies consist in plant materials (seeds, bulbs, leaves) or plant-derived products, including extracts obtained using a range of extraction procedures and different aqueous or organic solvents (ethanol, methanol, ethyl acetate, hexane, butane, acetone, benzene, petroleum ether, etc.), or other preparations such as essential oils, concoctions and decoctions (Bulfon et al. 2013). Herbs such as *S. trilobatum*, *A. paniculata* and *P. corylifolia* were found to reduce *Vibrio* in *P. monodon* three times when supplied in enriched *Artemia*. Several plant products found to have potent antiviral activity against fish and shrimp viruses.

Antifungal properties were also found in many plants. Herbal compounds have the ability to inhibit the generation of oxygen anions and scavenge free radical, hence reducing stress effects. Other herbs such as *Astragalus membranaceus*, *Portulaca oleracea*, *Sophora flavescens* and *A. paniculata* and many others are known to have specific and non-specific antistress effects.

Nowadays, only few commercial herbal products are available at a global level for large-scale use in aquaculture. In many countries a review of the current legislation should be undertaken to allow a greater flexibility in their use taking into consideration the benefits that they might have in intensive farming conditions, in terms of fish welfare and public health. Plants and plant bioactives might be proposed in aquaculture primarily as feed additives or immunostimulants, rather than therapeutics, as the registration of herbal remedies to be used in this field is a time-consuming process and implies higher economic costs (Bulfon et al. 2013).

Previously, studies have indicated that ginger (*Zingiber officinale*) and/or garlic (*Allium sativum*) is effective for the control of a range of bacterial, fungal and parasitic conditions. They also have an anti-inflammatory and anti-oxidative activity, as well as being effective as immunomodulatory agents in animals, including fish. Furthermore, they have been studied for their potential to control *Aeromonas hydrophila* infection in rainbow trout. Ginger and garlic are recognized to have broad-spectrum activity including activation of phagocytic cells, which is an important component of the non-specific immune system of fish.

The results of those studies reinforce the growing view that some plants are beneficial to fish by conferring protection against disease and stimulating the immune response (Nya and Austin 2009).

A study suggest that salinomycin with amprolium may be a promising treatment for myxosporean infections in intensively cultured warm-water fish, exhibiting action partially via the enhancement of host, innate immune functions and leading to parasite elimination (Karagouni et al. 2005).

Neem (*Azadirachta indica*) is effective and qualifies as safe and efficient in the prevention of ichthyophonosis in fish. Based on aforementioned results, the following conclusions could be recommended as the effective role of neem in the treatment of ichthyophonosis in *O. niloticus* fish since neem stimulated both humoral and cell-mediated immunity and succeeded for the first time to eradicate all the *Ichthyophonus* spores in fish after 3 months of treatments (Abd El-Ghany et al. 2008).

Marine organisms are potentially prolific sources of highly bioactive secondary metabolites that might represent useful leads in the development of new pharmaceutical agents. Antibacterial activity of methanolic extracts from 20 species of macroalgae (9 Chlorophyta, 3 Phaeophyta and 8 Rhodophyta) was evaluated against *Escherichia coli*, *Staphylococcus aureus* and *Enterococcus faecalis* (Zbakh et al. 2012).

The extracts of the studied 26 marine Rhodophyceae (8 Ceramiales, 7 Gelidiales, 9 Gigartinales, 1 Bonnemaisoniales and 1 Rhodymeniales) (Bouhlal et al. 2010) inhibited considerably the growth of the three tested bacterial strains and gave inhibition zones between 20 and 24 mm. *Staphylococcus aureus* was the most susceptible microorganism (10–35 mm of inhibition). The results indicate that these species of seaweed present a significant capacity of antibacterial activities, which makes them interesting for screening for natural products (Zbakh et al. 2012; Bouhlal et al. 2010).

Immunostimulants such as glucan, chitin, lactoferrin, levamisole and some medicinal plant extracts or products have been used to control fish and shellfish diseases. The immunostimulants mainly facilitate the function of phagocytic cells, increase their bactericidal activities and stimulate the natural killer cells, complement, lysozyme activity and antibody responses in fish and shellfish which confer enhanced protection from infectious diseases.

Administration of herbal extracts or their products at various concentrations through oral (diet) or injection route enhances the innate and adaptive immune response of different freshwater and marine fish and shellfish against bacterial, viral and parasitic diseases (Harikrishnan et al. 2011).

The development of nonantibiotic and environmentally friendly agents is one of the key factors for health management in aquaculture.

Consequently, with the emerging need for environmentally friendly aquaculture, the use of alternatives to antibiotics in fish nutrition is now widely accepted. In recent years, probiotics have taken centre stage and are to be used as an unconventional approach that has numerous beneficial effects in fish and shellfish culture: improved activity of gastrointestinal microbiota and enhanced immune status, disease resistance, survival, feed utilization and growth performance. As natural products, probiotics have much potential to increase the efficiency and sustainability of aquaculture production.

The concept of biological disease control, particularly using microbiological modulators for disease prevention, has received widespread attention. A bacterial supplement of a single or mixed culture of selected non-pathogenic bacterial strains is termed probiotics.

Probiotics thus are opening a new era in the health management strategy from human to aquatic species including fish and shellfish.

Probiotics were found to stimulate the feed conversion efficiency, augment live weight gain in fish and shrimp culture and confer protection against pathogens by competitive exclusion for adhesion sites, production of organic acids, hydrogen peroxide, antibiotics, bacteriocins, siderophores and lysozyme and also modulate physiological and immunological responses in fish. Moreover, probiotics are also being used as biological control agents in highly stocked intensive aquaculture ponds (Bidhan et al. 2014; Martinez Cruz et al. 2012; Lazado et al. 2014). A wide range of microalgae (*Tetraselmis*), yeast (*Debaryomyces*, *Phaffia* and *Saccharomyces*), Gram-positive bacteria (*Bacillus*, *Lactococcus*, *Micrococcus*, *Carnobacterium*, *Enterococcus*, *Lactobacillus*, *Streptococcus*, *Weissella*) and Gram-negative bacteria (*Aeromonas*, *Alteromonas*, *Photorhodobacterium*, *Pseudomonas* and *Vibrio*) has been evaluated as probiotics.

Several microalgae, yeasts and Gram-positive and Gram-negative bacteria have been isolated from the aquatic medium. Likewise, probiotics have been characterize as new eco-friendly alternative measures of disease control in aquaculture. Generally, probiotics have proven their promising growth results in fish by enhancing the feed conversion efficiency, as well as conferring protection against harmful bacteria by competitive exclusion, production of organic acids, hydrogen peroxide and several other compounds (Bidhan et al. 2014).

In various experiments, probiotics administered to tilapia (*Oreochromis niloticus*) increased non-specific immune response, determined by parameters such as lysozyme activity, neutrophil migration and bactericidal activity, which improved the resistance of fish to infection by *Edwardsiella tarda*. Other researchers isolated a strain of *Carnobacterium* sp. from salmon bowel and administered alive to rainbow trout and Atlantic salmon, demonstrating in vitro antagonism against known fish pathogens: *Aeromonas hydrophila*, *A. salmonicida*, *Flavobacterium psychrophilum*, *Photobacterium damselae* and *Vibrio* species. There is also evidence on the effect of dead probiotic cultures consisting on a mixture of *Vibrio fluvialis* A3-47S, *Aeromonas hydrophila* A3-51 and *Carnobacterium* BA211, in the control of furunculosis in rainbow trout.

For shrimp, studies have focused on the evaluation of probiotics such as *Bacillus cereus*, *Paenibacillus polymyxa* and *Pseudomonas* sp. PS-102 as biocontrol agents against pathogens of various *Vibrio* species.

Probiotic strains isolated from the gastrointestinal tract of clownfish (*Amphiprion percula*) have been used to inactivate several pathogens such as *Aeromonas hydrophila* and *Vibrio alginolyticus* among others. Probiotic promotes the development of healthy microbiota in the gastrointestinal tract of ornamental fishes from the genera *Poecilia* and *Xiphophorus*, decreasing the amount of heterotrophic microorganisms.

It was reported that the use of *Vibrio alginolyticus* strains as probiotics increases survival and growth of white shrimp.

Antibiotic are used, in some cases, as a part of conventional intensive animal farming and finfish aquaculture. Increased public concern about antibiotic resistance and the need to preserve the ever-diminishing arsenal of antimicrobials that work in humans for as long as possible have brought about increased scrutiny of the use of antibiotics – especially for prophylactic and growth-enhancing purposes. In accordance with European regulations and to limit the phenomenon of antibiotic resistance, studies are being implemented on the use of herbal or homeopathic medicine and probiotics, which are administered in addition to the feed.

A revision of the European Regulation relating to veterinary medicinal products is currently under way (proposal for a regulation of the European Parliament and of the Council – COM 2014 558 final). In this proposal, article 4 states:

> "Veterinary medicinal product" means any substance or combination of substances which fulfils at least one of the following conditions:
>
> (a) it is presented as having properties for treating or preventing disease in animals;
>
> (b) its purpose is to be used in or administered to animals with a view to restoring, correcting or modifying physiological functions by exerting a pharmacological, immunological or metabolic action, or to making a medical diagnosis;
>
> (c) its purpose is to be used for euthanasia of animals.
>
> "Substance" means any matter of the following origin: (a) human, (b) animal, (c) vegetable, (d) chemical.
>
> "Homeopathic veterinary medicinal product" means a veterinary medicinal product prepared from homeopathic stocks in accordance with a homeopathic manufacturing procedure described by the European Pharmacopoeia or, in the absence thereof, by the pharmacopoeias used officially in Member States.

In recent years there is increasing experimental evidence and studies of probiotics and herbal medicine, and the first results seem to confirm their effectiveness in the prevention and management of diseases affecting aquatic animals breeding (Bulfon et al. 2016).

The use of these substances is permitted in accordance with article 25(t) of Regulation 889/2008 but does not describe in what way and in what quantities are to be administered and they are authorized. It would be appropriate to make a list of such microorganisms and plants which can be used in the composition of the feed, for example, as shown in the register of animal feed additives of the Annex to Regulation 2003/1831 (* extracts and microorganisms).

The extracts of the following plants have been tested to prove their effectiveness against diseases that primarily affect livestock, particularly if they are effective against bacteria, such as *Aeromonas* sp. and *Vibrio* sp., other microorganisms, viruses, fungi and parasites. The plants tested were *Solanum trilobatum*, *Andrographis paniculata (*)*, *Psoralea corylifolia*, *Astragalus membranaceus (*)*, *Portulaca oleracea*, *Sophora flavescens*, *Zingiber officinale (*)*, *Allium sativum*, *Origanum vulgare (*)*, *Azadirachta indica (*)*, *marine algae*, *Rhodophyceae*, *Achyranthes aspera*, *Angelica sinensis (*)*, *Cynodon dactylon*, *Echinacea purpurea (*)*, *Massa medicated*, *Punica granatum (*)*, *Solanum nigrum*, *Withania somnifera (*)* and *Zataria multiflora.*

The probiotics most tested and which have given the best results in the trials were microalgae (*Tetraselmis*), yeasts (*Debaryomyces*, *Phaffia*, *Saccharomyces*), Gram-positive bacteria (*Bacillus* (*), *Lactococcus* (*), *Micrococcus*, *Carnobacterium*, *Enterococcus* (*), *Pediococcus* (*), *Lactobacillus* (*), *Streptococcus* (*), *Weissella*) and Gram-negative bacteria (*Aeromonas*, *Alteromonas*, *Pseudomonas*, *Vibrio*).

There are initial investigations and tests with regard to the preparation of vaccines derived from the study of genetic engineering, such as DNA-recombinant vaccines (Regulation 2003/1829 article16), and proteins produced from GMOs. From the first studies, we can see how it is possible to produce new solutions for disease prevention obtaining vaccines and immunostimulants with low-cost and low environmental impact. It would be interesting to continue to do studies and tests in this direction, since the Regulation 834/2007 Article 4 allows for the use of GMOs to produce veterinary medicinal products.

Biosecurity

Biosecurity consists of practices that minimize the risk of spreading an infectious disease among the animals within a facility, but also the risk that diseased animals or infectious agents are spread to other sites, to other susceptible species or can contaminate the environment. These practices also reduce stress to the animals, thus making them less susceptible to disease. Good biosecurity measures will reduce the risk of relevant losses from infectious disease and low-level losses that, over time, can also greatly affect the bottom line.

A comprehensive biosecurity plan in aquaculture farming should include the following activities:

- Monitoring system of water quality and fish health.
- Obtaining healthy animals from a reputable supplier.
- Good preventive practices like quarantine, routine observation and vaccination, access control, foot baths and handwashing in critical facilities (hatchery and nursery).
- Isolating sick fish and removing dead and moribund fish.
- Updated knowledge about diagnostics to be used and which treatments are legal and available.
- Good husbandry and feeding practices to reduce stress.
- Daily maintenance and disinfection of equipment and disinfection of trucks/vehicles accessing the farm.
- Always check product characteristics for appropriate concentrations, use, shelf life and safety precautions.
- Training of personnel and information to visitors so that they understand and follow biosecurity protocols.
- Keeping good records and documentation of biosecurity protocols.
- When necessary, consult with the state environmental control agency or the hazardous waste representative at the nearest EPA regional office for guidance on the proper disposal of each product (Oidtmann et al. 2011; Galli et al. 2014; Pietrak et al. 2014; Manual of Diagnostic Tests for Aquatic Animals OIE 2009; Southern Regional Aquaculture Centre (SRAC) 2013).

Effective biosecurity plans must be tailored to a specific farm site, be adaptable, address local disease threats and avoid environmental insult. The biosecurity policies and practices of an aquaculture company are controlled directly by the farmer. The goals of these policies and practices match those of the various levels of government-regulated biosecurity, i.e. to reduce the probability that a pathogen will infect one or more animals under the farmer's care or negatively impact the surrounding farms or environment.

A good biosecurity plan, consistently implemented, functions as a type of insurance policy against disease. The routine use of biosecurity measures (secure water supply, healthy fish or shellfish stock, good hygiene practices for all entering and exiting the farm) can reduce the risk of introduction and economic impact of these diseases on the farm (Pietrak et al. 2014).

The washing and disinfection procedures should at least include the following stages:

(a) Removal of solid waste, etc. followed by prewashing
(b) Deep cleaning and washing
(c) Disinfection
(d) Rinsing

The process should be monitored throughout by a technically competent person, and records need to be kept (Manual of Diagnostic Tests for Aquatic Animals OIE 2017).

For carp breeding, when hygienic measures are necessary, quick lime (CaO) is permitted to be applied onto the humid pond bottom (max. 200 kg ha^{-1}). Its application into the pond (max. 150 kg ha^{-1}) for the purposes of pH stabilization and for precipitating of suspended organic matter is permitted in critical weather situations (Adámek et al. 2014; Horváth et al. 2015).

For the culture of Salmonidae, Coregonidae, Gadidae and croakers/drum for controlling sea lice in marine net cages, stocking with wrasse as "cleaner fishes" is recommended; for the protection of net cages against growth of algae and colonization by invertebrates, environment-friendly methods shall be employed.

Health status of animals shall be monitored and documented on a regular basis. Special efforts shall be made to detect correlation between management measures, manifestation of viral diseases, causes of mortality, individual growth and yields/biomass development.

Good hygiene practices and farm management prevent the onset of diseases. General guidelines are provided by the OIE-World Organization for Animal Health Aquatic Animal Health Code, 2017, but a specific biosecurity protocol, to be used in the most important species farmed in Europe, is available only at national level at the moment. It would be appropriate in the next future to share at European level the same good hygiene practices in compliance with Council Directive 2006/88 and the new Regulation EU 2016/429 on transmissible animal diseases and amending and repealing certain acts in the area of animal health ("Animal Health Law").

References

Abd El-Ghany NA, Abd Alla HML (2008) A trial for treatment of Ichthyophonosis in cultured *Oreochromis niloticus* using focus and neem plants. 8th international symposium on Tilapia in aquaculture. Cairo, Egypt

Adámek Z, Helešic J, Maršálek B, Rulík M (2014) Applied hydrobiology. FROV JU Vodňany, 376 p. http://www.rybarskeknihy.cz/48-english-books

Ashley PJ (2007) Fish welfare: current issues in aquaculture. Appl Anim Behav Sci 104(3–4):199–235

Bagni M, Civitareale C, Priori A, Ballerini A, Finoia MG, Brambilla G, Marino G (2007) Pre-slaughter crowding stress and killing procedures affecting quality and welfare in sea bass (*Dicentrarchus labrax*) and sea bream (*Sparus aurata*). Aquaculture 263:52–60

Barnabé G (1990) Rearing bass and gilthead bream. In: Barnabé G (ed) Aquaculture. Ellis Horwood, New York, pp 647–686

Barnabé G (1991) Grossissement des poissons en élevage intensif. In: Barnabé G (ed) Bases biologiques et écologiques de l'aquaculture. Lavoisier-Tec & Doc, Paris, pp 422–451

Barton BA (2002) Stress in fishes: a diversity of responses with particular reference to changes in circulating corticosteroids. Integ Comp Biol 42:517–525

Batzina A, Dalla C, Papadopoulou-Daifoti Z, Karakatsouli N (2014) Effects of environmental enrichment on growth, aggressive behaviour and brain monoamines of gilthead seabream *Sparus aurata* reared under different social conditions. Comp Biochem Physiol A Mol Integr Physiol 169:25–32

Berg T, Erikson U, nordvedt TS (1997) Rigor mortis assessment of Atlantic Salmon (*Salmo salar*) and effects of stress. J Food Sci 62(3):439–446

Bidhan De C, Meena DK, Behera BK, Das P, Das Mohapatra PK, Sharma AP (2014) Probiotics in fish and shellfish culture: immunomodulatory and ecophysiological responses. Fish Physiol Biochem 40(3):921–971

Bouhlal R, Riadi H, Martínez J, Bourgougnon N (2010) The antibacterial potential of the sea-weeds (Rhodophyceae) of the Strait of Gibraltar and the Mediterranean coast of Morocco. Afr J Biotechnol 9(38):6365–6372

Bulfon C, Bongiorno T, Messina M, Volpatti D, Tibaldi E, Tulli F (2016) Effects of anax gin-seng extract in practical diets for rainbow trout (*oncorhynchus mykiss*) on growth performance, immune response and resistance to *ersinia ruckeri*. Aquac Res 48(5):2369–2379

Bulfon C, Volpatti D, Galeotti M (2013). Current research on the use of plant-derived products in farmed fish. Aquac Res 46(3). https://doi.org/10.1111/are.12238

Bulfon C, Volpatti D, Galeotti M (2014) In vitro antibacterial activity of plant ethanolic extracts against fish pathogens. J World Aquac Soc 45(5). https://doi.org/10.1111/jwas.12151

Carbonara P, Corsi I, Focardi S, Lembo G, Rochira S, Scolamacchia M, Spedicato MT, Mckinley RS (2010) The effects of stress induced by cortisol administration on the repeatability of swim-ming performance tests in the European sea bass (*Dicentrarchus labrax L.*). Mar Freshw Behav Physiol 43:283–296

Carbonara P, Scolamacchia M, Spedicato MT, Zupa W, Mckinley RS, Lembo G (2013) Muscle activity as a key indicator of welfare in farmed European sea bass (*Dicentrarchus labrax, L.* 1758). Aquac Res 2014:1–14

Chandroo KP, Duncan IJH, Moccia RD (2004) Can fish suffer?: Perspectives on sentience, pain, fear and stress. Appl Anim Behav Sci 86:225–250

Chatelier A, McKenzie DJ, Prinet A, Galois R, Robin J, Zambonino J, Claireaux G (2006) Associations between tissue, fatty acid composition and physiological traits of performance in the sea bass (*Dicentrarchus labrax*). J Exp Biol 209:3429–3439

Claireaux G (2006) Effect of temperature on maximum swimming speed and cost of transport in juvenile European sea bass (*Dicentrarchus labrax*). J Exp Biol 209(17):3420–3428

Claireaux G, Lagardère J (1999) Influence of temperature, oxygen and salinity on the metabolism of the European sea bass. J Sea Res 42:157–168

Claireaux G, Webber DM, Lagardère J-P, Kerr SR (2000) Influence of water temperature and oxygenation on the aerobic metabolic scope of Atlantic cod (*Gadus morhua*). J Sea Res 44(3–4):257–265

Conte FS (2004) Stress and the welfare of cultured fish. Appl Anim Behav Sci 86:205–223

Council of Europe – Standing Committee of the European Convention for The Protection Of Animals Kept For Farming Purposes (T-Ap) Recommendation concerning farmed fish adopted by the Standing Committee on 5 December 2005

d'Orbcastel ER, Lemarié G, Breuil G, Petochi T, Marino G, Triplet S, Dutto G, Fivelstad S, Coeurdacier J-L, Blancheton J-P (2010) Effects of rearing density on sea bass (*Dicentrarchus labrax*) biological performance, blood parameters and disease resistance in a flow through system. Aquat Living Resour 23(1):109–117

Dawkins MS (1990) From an animal's point of view: motivation, fitness, and animal welfare. Behav Brain Sci 13:1–61

Delince GA, Campbell D, Janssen JAL, Kutty MN (1987) FAO: seed production. PROJECT RAF/82/009

Duncan IJH (2005) Science-based assessment of animal welfare: farm animals. Rev Sci Tech Off Int Epiz 24:483–492

Duncan IJH, Dawkins MS (1983) The problem of assessing 'well-being' and 'suffering' in farm animals. In: Smidt D (ed) Indicators relevant to farm animal welfare. Martinus Nijhoff, The Hague, pp 13–24

EFSA (2008) Animal welfare aspects of husbandry systems for farmed European seabass and gilthead seabream. EFSA J 844:1–21

EFSA (2009a) Food Safety considerations concerning the species specific welfare aspects of the main systems of stunning and killing of farmed fish. EFSA J 1190:1–16

EFSA (2009b) Species-specific welfare aspects of the main systems of stunning and killing of farmed seabass and seabream. EFSA J 1010:1–52

Ellis T, North B, Scott AP, Bromage NR, Porter M, Gadd D (2002) The relationships between stocking density and welfare in farmed rainbow trout. J Fish Biol. https://doi.org/10.1111/j.1095-8649.2002.tb00893.x

FSBI (2002) Fish welfare. Briefing paper 2

Galli L, Griffiths D, Jiravanichpaisal P, Wattanapongchart N, Wongsrirattanakul O, Shinn A (2014) Biosecurity in aquaculture part 1: international considerations. The fish site www.thefishsite.com/articles

Giuffrida A, Pennisi L, Ziino G, Fortino L, Valvo G, Marino S, Panebianco A (2007) Influence of slaughtering method on some aspects of quality of gilthead seabream and smoked rainbow trout. Vet Res Commun 31(4):437–446

Glaropoulos A, Papadakis VM, Papadakis IE, Kentouri M (2012) Escape-related behavior and coping ability of sea bream due to food supply. Aquac Int 20(5):965–979

Harikrishnan R, Balasundaram C, Heo M-S (2011) Impact of plant products on innate and adaptive immune system of cultured finfish and shellfish. Aquaculture 317(1–4):1–15

Heath AG (1995) Water pollution and fish physiology. VPI & State University, Blacksburg, 359 pp

Horváth L, Tamás G, Coche AG, Kovács E, Moth-Poulsen T, Woynarovich A (2015) Training manual on the advanced fry and fingerling production of carp in ponds. FAO REU, Budapest. http://www.fao.org/3/a-i4317e.pdf

Houston AH (1990) Blood and circulation. In: Schreck CB, Moyle PB (eds) Methods for fish biology. American Fisheries Society, Maryland, pp 273–334

Huntingford FA (2004) Implication of domestication and rearing conditions for the behaviour of cultivated fish. J Fish Biol 65:122–142

Huntingford FA, Adams C, Braithwaite VA, Kadri S, Pottinger TG, Sandoe P, Turnbull JF (2006) Current issues in fish welfare. J Fish Biol 68(2):332–372

Huntingford FA, Kadri S (2009) Taking account of fish welfare: lessons from aquaculture. J Fish Biol 75(10):2862–2867

Huntingford FA, Kadri S (2008) Welfare and fish. Wiley online library. https://doi.org/10.1002/9780470697610.ch2

Hvas M, Folkedal O, Imsland A, Oppedal F (2017) The effect of thermal acclimation on aerobic scope and critical swimming speed in Atlantic salmon. J Exp Biol 220(15):2757–2764

Iversen MH, Eliassen RA (2009) The effect of AQUI-S® sedation on primary, secondary, and tertiary stress responses during salmon smolt (*Salmo salar L.*) transport and transfer to sea. J World Aquacult Soc 40(2):216–225. https://doi.org/10.1111/j.1749-7345.2009.00244.x

Iversen M, Eliassen R, Bengt F (2008) Potential benefit of clove oil sedation on animal welfare during salmon smolt, *Salmo salar L.* transport and transfer to sea. Aquac Res 40(2):233–241. https://doi.org/10.1111/j.1365-2109.2008.02091.x

Karagouni E, Athanassopoulou F, Lytra A, Komis C, Dotsika E (2005) Antiparasitic and immunomodulatory effect of innovative treatments against Myxobolus sp. infection in *Diplodus puntazzo*. Vet Parasitol 134(3–4):215–228

Kolkovski S, Kolkovski J (2011) Herbal medicine in aquaculture. Int Aquafeed 14(2):28–31

Korte SM, Olivier B, Koolhaas JM (2007) A new animal welfare concept based on allostasis. Physiol Behav. https://doi.org/10.1016/j.physbeh.2006.10.018

Lazado CC, Caipang CMA (2014) Atlantic cod in the dynamic probiotics research in aquaculture. Aquaculture 424–425:53–62

Lembo G, Zupa W (2010) Fish welfare, a key issue for organic system standards. In the IFOAM Dossier – Organic Aquaculture. Background, Assessment, Interpretation. http://www.ifoam-eu.org/en/library/dossiers

Lembo G, Carbonara P, Scolamacchia M, Spedicato MT, Mckinley RS (2007) Use of muscle activity indices as a relative measure of well-being in cultured sea bass *Dicentrarchus labrax* (Linnaeus, 1758). Hydrobiologia 582:271–280

Lambooij E, Grimsbø E, van de Vis JW, Reimert HGM, Nortvedt R, Roth B (2010) Percussion and electrical stunning of Atlantic salmon (*Salmo salar*) after dewatering and subsequent effect on brain and heart activities. Aquaculture 300(1–4):107–112

Lines JA, Spence J (2012) Safeguarding the welfare of farmed fish at harvest. Fish Physiol Biochem 38(1):153–162

Losordo TM, Rakocy J, Masser M (1999) Recirculating aquaculture tank production systems: a review of component options. South Reg Aquac Cent 453:12

Lowe TE, Ryder JM, Carragher JF, Wells RMG (1993) Flesh quality in snapper, *Pagrus auratus,* affected by capture stress. J Food Sci 58:770–773

Lucas MC (1994) Heart rate as an indicator of metabolic rate and activity in adult Atlantic salmon, *Salmo salar*. J Fish Biol 44:889–903. https://doi.org/10.1111/j.1095-8649.1994.tb01262.x

Marine Harvest (2014). Salmon farming industry handbook. www.marineharvest.com/investor/industry-handbook/

Martins CIM, Galhardo L, Noble C, Damsgard B, Spedicato MT, Zupa W, Beauchaud M, Kulczykowska E, Massabuau J-C, Carter T, Planellas SR, Kristiansen T (2012) Behavioural indicators of welfare in farmed fish. Fish Physiol Biochem 38:17–41

Martínez Cruz P, Ibáñez AL, Monroy Hermosillo OA, Ramírez Saad HC (2012) Use of probiotics in aquaculture. ISRN Microbiol 2012:1–13

Masser MP, Rakocy J, Losordo TM (1999) Recirculating aquaculture tank production systems: management of recirculating systems. South Reg Aquac Center 452:12

McKenzie DJ, Pedersen PB, Jokumsen A (2007) Aspects of respiratory physiology and energetics in rainbow trout (*Oncorhynchus mykiss*) families with different size-at-age and condition factor. Aquaculture 263(1–4):280–294

Mellor DJ, Stafford KJ (2001) Integrating practical, regulatory and ethical strategies for enhancing farm animal welfare. Aust Vet J 79(11):762–768

Mommsen TP, Vijayan MM, Moon TW (1999) Cortisol in teleost: dynamics, mechanisms of action, and metabolic regulation. Rev Fish Biol Fish 9:211–268

Montero D, Izquierdo MS, Tort L, Robaina L, Vergara JM (1999) Fish Physiol Biochem 20(1):53–60

Mylonas CC, Cardinaletti G, Sigelaki I, Polzonetti-Magni A (2005) Comparative efficacy of clove oil and 2-phenoxyethanol as anesthetics in the aquaculture of European sea bass (*Dicentrarchus labrax*) and gilthead sea bream (*Sparus aurata*) at different temperatures. Aquaculture 246:467–481

Nakayama T, Ooguchi N, Ooi A (1999) Change in rigor mortis of red sea bream dependent on season and killing method. Fish Sci 65(2):284–290

Norin T, Clark TD (2016) Measurement and relevance of maximum metabolic rate in fishes. J Fish Biol 88(1):122–151

Nya EJ, Austin B (2009) Use of dietary ginger, *Zingiber officinale* Roscoe, as an immunostimulant to control *Aeromonas hydrophila* infections in rainbow trout, *Oncorhynchus mykiss* (Walbaum). J Fish Dis 32:971–977. https://doi.org/10.1111/j.1365-2761.2009.01101.x

Oidtmann BC, Thrush MA, Denham KL, Peeler EJ (2011) International and national biosecurity strategies in aquatic animal health. Aquaculture 320(1–2):22–33

Oppedal F, Dempster T, Stien LH (2011) Environmental drivers of Atlantic salmon behaviour in sea-cages: a review. Aquaculture 311(1–4):1–18

Papadakis IE, Papadakis VM, Glaropoulos A, Petroutsos SI, Gonzalez L, Kentouri M (2013) Do the visual conditions at the point of escape affect European sea bass escape behavior?. Mediterranean Marine Science. https://doi.org/10.12681/mms.360

Pedersen CL (1987) Energy budgets for juvenile rainbow trout at various oxygen concentrations. Aquaculture 62(3–4):289–298

Person-Le Ruyet J, Le Bayon N (2009) Effects of temperature, stocking density and farming conditions on fin damage in European sea bass (*Dicentrarchus labrax*). Aquat Living Resour 22:349–362

Person-Le Ruyet J, Mahé K, Le Bayon N, Le Delliou H (2004) Effects of temperature on growth and metabolism in a Mediterranean population of European sea bass, *Dicentrarchus labrax*. Aquaculture 237:269–280

Pickering AD, Pottinger TG, Sumpter JP, Carragher JF, Le Bail PY (1991) Effects of acute and chronic stress on the levels of circulating growth hormone in the rainbow trout, *Oncorhynchus mykiss*. Gen Comp Endocrinol. https://doi.org/10.1016/0016-6480(91)90108-I

Pietrak M, Leavitt D, Walsh M (2014) Biosecurity on the farm – guidelines & resources for developing a biosecurity plan. The fish site www.thefishsite.com/articles

Poli BM, Parisi G, Scappini F, Zampacavallo G (2005) Fish welfare and quality as affected by pre-slaughter and slaughter management. Aquac Int 13:29–49

Robb DHF, Roth B (2003) Brain activity of Atlantic salmon (*Salmo salar*) following electrical stunning using various field strengths and pulse durations. Aquaculture 216(1–4):363–369

Roth B, Birkeland S, Oyarzun F (2009) Stunning, pre slaughter and filleting conditions of Atlantic salmon and subsequent effect on flesh quality on fresh and smoked fillets. Aquaculture 289(3–4):350–356

Roth B, Imsland A, Moeller D, Slinde E (2003) Effect of electric field strength and current duration on stunning and injuries in market-sized Atlantic Salmon held in seawater. N Am J Aquac 65(1):8–13

Roth B, Moeller D, Slinde E (2004) Ability of electric field strength, frequency, and current duration to stun farmed Atlantic Salmon and Pollock and relations to observed injuries using sinusoidal and square wave alternating current. N Am J Aquac 66(3):208–216

Roncarati A, Melotti P, Dees A, Mordenti O, Angelotti L (2006) Welfare status of cultured seabass (*Dicentrarchus labrax L.*) and seabream (*Sparus aurata* L.) assessed by blood parameters and tissue characteristics. J App Ichthyol 22:225–234. https://doi.org/10.1111/j.1439-0426.2006.00741.x

Saillant E, Fostier A, Haffray P, Laureau S, Thimonier J, Chatain B (2003a) Effects of rearing density, size grading and parental factors on sex ratios of the sea bass (*Dicentrarchus labrax L.*) in intensive aquaculture. Aquaculture 221:183–206

Saillant E, Fostier A, Haffray P, Menu B, Chatain B (2003b) Saline preferendum for the European sea bass, *Dicentrarchus labrax*, larvae and juveniles: effect of salinity on early development and sex determination. J Exp Mar Biol Ecol 287:103–117

Santos GA, Schrama JW, Mamauag REP, Rombout JHWM, Verreth JAJ (2010) Chronic stress impairs performance, energy metabolism and welfare indicators in European seabass (*Dicentrarchus labrax*): the combined effects of fish crowding and water quality deterioration. Aquaculture 299(1–4):73–80

Saroglia M, Terova G, De Stradis A, Caputo A (2002) Morphometric adaptations of sea bass gills to different dissolved oxygen partial pressures. J Fish Biol 60:1423–1430

Schram E, Roques JAC, Abbink W, Spanings T, de Vries P, Bierman S, van de Vis H, Flik G (2010) The impact of elevated water ammonia concentration on physiology, growth and feed intake of African catfish (*Clarias gariepinus*). Aquaculture 306(1–4):108–115

Skov PV, Larsen BK, Frisk M, Jokumsen A (2011) Effects of rearing density and water current on the respiratory physiology and haematology in rainbow trout, *Oncorhynchus mykiss* at high temperature. Aquaculture 319(3–4):446–452

Turnbull JF, North BP, Ellis T, Adams CE, Bron J, Mac Intyre CM, Huntingford FA (2008) Stocking density and the welfare of farmed salmonids. Wiley online library. https://doi.org/10.1002/9780470697610.ch8

Urbieta FJ, Ginés R (2000) Optimisation of slaughtering method in gilthead seabream (*Sparus aurata*). Industrial application in fish farms. Cah Opt Méd 51:71–77

Van de Vis H, Abbink W, Lambooij B, Bracke M (2014) Stunning and killing of farmed fish: how to put it into practice? In: Devine C, Dikeman M (eds) Encyclopaedia of meat sciences, vol 3, 2nd edn. Elsevier, Oxford, pp 421–426

Wajsbrot N, Gasith A, Krom MD, Popper DM (1991) Acute toxicity of ammonia to juvenile gilthead seabream *Sparus aurata* under reduced oxygen levels. Aquaculture 92:277–288

Wedemeyer GA (1996) Physiology of fish in intensive culture systems. Chapman & Hall, New York. 232 pp

Ytteborg E, Baeverfjord G, Torgersen J, Hjelde K, Takle H (2010) Molecular pathology of vertebral deformities in hyperthermic Atlantic salmon (*Salmo salar*). BMC Physiol 10(1):12

Zampacavallo G, Parisi G, Mecatti M, Lupi P, Giorgi G, Poli BM (2015) Evaluation of different methods of stunning/killing sea bass (*Dicentrarchus labrax*) by tissue stress/quality indicators. J Food Sci Technol 52(5):2585–2597. https://doi.org/10.1007/s13197-014-1324-8

Zbakh H, Chiheb H, Bouziane H, Sánchez VM, Riadi H (2012) Antibacterial activity of benthic marine algae extracts from the Mediterranean coast of Morocco. J Microbiol Biotechnol Food Sci 2(1):219–228

Zupa W, Carbonara P, Spedicato MT, Lembo G (2015) Modelling swimming activities and energetic costs in European sea bass (*Dicentrarchus labrax* L., 1758) during critical swimming tests. Mar Freshw Behav Physiol 48(5):341–357. https://doi.org/10.1080/10236244.2015.1073456

Chapter 8
Nutrition in Relation to Organic Aquaculture: Sources and Strategies

Elena Mente, Alfred Jokumsen, Chris G. Carter, Efi Antonopoulou, and Albert G. J. Tacon

Introduction

Organic production is a system of farm management and food production that combines best environmental practices, a high level of biodiversity, the preservation of natural resources, the application of high animal welfare standards and a production method in line with the preference of certain consumers for products produced using natural substances and processes. Mie et al. (2017) reviewed existing evidence on the impact of organic food on human health and compared organic versus conventional food production with respect to parameters important to human health. The review emphasised several documented human health benefits associated with organic food production and production methods and concluded that it is likely to

E. Mente (✉)
Department of Ichthyology and Aquatic Environment, School of Agricultural Sciences, University of Thessaly, Volos, Greece
e-mail: emente@uth.gr

A. Jokumsen
National Institute of Aquatic Resources, (DTU Aqua), Technical University of Denmark, Nordsøen Forskerpark, Hirtshals, Denmark
e-mail: ajo@aqua.dtu.dk

C. G. Carter
Institute for Marine and Antarctic Studies, University of Tasmania, Hobart, Tasmania, Australia
e-mail: chris.carter@utas.edu.au

E. Antonopoulou
Laboratory of Animal Physiology, Department of Zoology, School of Biology, Faculty of Sciences, Aristotle University of Thessaloniki, University Campus, Thessaloniki, Greece
e-mail: eantono@bio.auth.gr

A. G. J. Tacon
Aquatic Farms Ltd, Kaneohe, HI, USA
e-mail: agjtacon@aquahana.com

G. Lembo, E. Mente (eds.), *Organic Aquaculture*,
https://doi.org/10.1007/978-3-030-05603-2_8

be beneficial within the conventional agriculture, for example, in integrated pest management and antibiotics. This chapter covers aspects of current use of formulated feeds, feed composition, aquafeed technology, sustainable alternatives to common feed ingredients, nutritional physiology and general nutritional principles and product quality in the context of the organic aquaculture. It reviews new knowledge and presents research results to update and may modify the criteria and standards for organic aquaculture in relation to nutrition and thus to provide high-quality products for the consumers. This chapter is based on the current European regulation on organic aquaculture, as well as on the proposed revision of the European regulation, which is currently being approved after a long process for getting the agreement of the European Parliament, European Council and the European Commission.

Feeds in Organic Aquaculture

Organic aquaculture reflects a specific production approach (Cottee and Petersan 2009) driven by the growing interest in sustainable utilisation of resources (Mente et al. 2011, 2012). The discussion and the debate on organic feeds for organic aquaculture is still open due to the balance that needs to be achieved between the fundamental rules in organic culture and the reality of the supply of the feed sources for aquafeeds. Nutrition of organically farmed aquatic animals implies that feeding shall be performed in a way that allows natural food intake and ensures that the developmental, physiological and behavioural needs of animals are met (KRAV 2009; Soil Association 2009; EC 2007, 2009, 2014). In addition, feeds must be balanced according to the nutritional requirements of the farmed organisms, promote animal's growth and health, ensure high quality of the final edible product and cause low environmental impact (KRAV 2009, 2010; Soil Association 2009; EC 2007, 2009, 2014).

Aquatic animals that are cultured in inland waters (i.e. ponds and lakes) should be provided by food materials, such as aquatic plants, algae, plankton, small invertebrates, detritus, etc., that are naturally available in the culture media (EC 2009). In semi-intensive production systems, where higher nutrient availability is required, the natural food productivity of the cultivated water can be enhanced by external inputs such as fertilisers, both of inorganic and organic nature (i.e. livestock manures, plant material and inorganic phosphate, nitrogen and potassium products), but they need to be certified as organic. If supplementary feeds as natural or naturally derived substances are offered in the above systems, they should be well documented, and evidence of the need to use them as an external input will be needed (EC 2007, 2009). In the case of omnivorous-carnivorous species cultured in inland waters, such as penaeid shrimps and freshwater prawns (*Macrobrachium* spp.), the EU production rule for organic aquaculture has set that their ration of supplementary organic feed may comprise a maximum of 25% of fishmeal and 10% of fish oil derived from sustainable fisheries (EC 2009, 2013, 2014). The feed ration of Siamese catfish (*Pangasius* spp.) may consist of a maximum of 10% fishmeal or fish oil derived from sustainable fisheries. In addition, with regard to bivalve molluscs, which are filter-feeding animals for the European organic aquaculture regulation and other species which are not fed by humans but instead feed on natural plankton, the following rules shall apply: (a) they shall receive all their nutritional requirements from nature, except in the case of juveniles reared in hatcheries and nurseries; and (b) the growing areas shall be suitable from a health point of view and shall either be of high ecological status as defined by Directive 2000/60/EC or of good environmental status as defined by Directive 2008/56/EC or of equivalent quality to the production zones classed as A in Regulation (EC) No. 854/2004, until 13 December 2019, or the corresponding classification areas set out in the implementing acts adopted by the Commission in accordance with Article 18(8) of Regulation (EU) 2017/625, from 14 December 2019.

In intensive aquaculture systems, feed is provided in the form of pellets that meet the animal's nutritional requirements at the various stages of its development (Mente et al. 2011). There is increasing concern about the consumption of fishmeal and fish oil for aquaculture feed due to the increasing demand from the expanding aquaculture industry and concerns about decreasing wild stocks. The use of fishmeal and fish oil contradicts to the organic principle of sustainability due to the decline of fisheries and overexploitation of wild stocks though; it is now possible for salmon farming to be a net producer of fish protein and oil (Crampton et al. 2010). Genetically modified organisms (GMOs) and products produced from or by GMOs as well as growth promoters are not in line with the concept of organic production and consumers' perception of organic products, EC 834/2007. Organic cultivated seaweed or sustainably harvested wild seaweed, including all multicellular marine algae or phytoplankton and microalgae, may be used as feed ingredients. Synthetic feed ingredients are not allowed, except feed additives, such as vitamin and mineral supplements, that are identical to natural and essential for nutritional purposes. The use of such additives should be used to a minimum extent and should not exceed requirements of the specific species. Synthetic antioxidants are not allowed, and only natural antioxidant substances should be used. Feed ingredients of mineral origin, trace elements, vitamins or provitamins shall be of natural origin. In case these substances are unavailable, chemically well-defined analogic substances may be authorised for use in organic production, EC 834/2007.

Feed for carnivorous organic aquaculture animals shall be sourced with the following priorities according to the EU regulation: (a) organic feed products of aquaculture origin; (b) fishmeal and fish oil from organic aquaculture trimmings; (c) fishmeal and fish oil and ingredients of fish origin derived from trimmings of fish already caught for human consumption in sustainable fisheries; (d) organic feed materials of plant or animal origin and especially plant material shall not exceed 60% of total ingredients; and (e) fishmeal and fish oil and ingredients of fish origin derived from whole fish, crustaceans or molluscs caught in fisheries certified as sustainable under a scheme recognised by the competent authority in line with the principles laid down in the EU Regulation 1380/2013 of the European Parliament and of the Council Regulation (EC) 889/2008 and not used for human consumption. Furthermore, the organic regulation does not allow balancing the dietary amino acid profile by supplementing with synthetic free amino acids to fulfil the dietary requirements of the specific organically produced species. However, the amino acid histidine produced through fermentation may be used in the feed for salmonid fish in case available feed sources do not provide a sufficient amount of histidine to meet the dietary needs of the fish and prevent the formation of cataracts, EC 1358/2014. The plant fraction of the feed shall originate from organic production and the feed fraction derived from aquatic animals shall originate from sustainable exploitation of fisheries. Nonorganic feed materials from plant origin, feed materials from animal and mineral origin, feed additives, certain products used in animal nutrition and processing aids shall be used only if they have been authorised for use in organic production under EC 834/2007. A summary of the acceptable and not acceptable materials in feeds for organic aquaculture according to the EU regulation is presented in Table 8.1.

All rules in relation to the nutrition of organically farmed aquatic animals suggest that the organic diet will meet the nutritional requirements of the farmed aquatic animal and will be offered to them in a way that allows natural feeding behaviour, with minimum loss of feed to the environment. In addition, the feed is comprised of natural products, in situ nutrient sources or organically produced products and by-products from organic food processing and waste products from the fisheries. Table 8.2 summarises the main similarities and differences to specific rules that apply to the feeding practices of aquatic animals reared in organic aquaculture as they have been defined in various international standards.

Alternatives Feed Ingredients to Overcome Bottlenecks in Organic Aquaculture

The demand to identify alternative sources of dietary protein and lipids for organic feeds in organic aquaculture and to reduce the use of fishmeal and fish oil in organic feeds is an ongoing effort due to the fact that wild fisheries are stagnated. However, the alternative ingredient quality and the ingredient certification to be used in organic aquaculture require increased attention. Research continues to evaluate novel formulated alternatives ingredients and assess product quality to meet the challenges for the production of the organic feeds.

Table 8.1 Acceptable and nonacceptable materials in feeds for organic aquaculture

Acceptable Materials
1. Feed materials of animal origin
1.1 Fish, other marine animals, their products and by-products
Under the following restrictions: products origin only from sustainable exploitation of fisheries resources under Common Fisheries Policy and to be used only for species other than herbivores
• Fish
• Fish oil and cod liver oil not refined
• Fish molluscan or crustacean autolysates
• Hydrolysate and proteolysates obtained by an enzyme action, whether in soluble form, solely provided to aquaculture animals
• Fishmeal
• Crustacean meal
• Fishmeal and fish oil ingredients of fish origin derived from trimmings of fish, crustaceans or molluscs already caught for human consumption
• Fishmeal and fish oil ingredients of fish origin derived from whole fish, crustaceans or molluscs caught in sustainable fisheries and not used for human consumption
• Feed products derived from whole fish caught in fisheries certified as sustainable under a scheme recognised by the competent authority in line with the principles laid down in Regulation (EU) No. 1380/2013 of the European Parliament and of the Council Regulation (EC) no. 889/2008, art. 25k (1)
Under the following restrictions: authorised under the above regulations for use in organic production and intended only for feeds of other farmed species or taxa
• Zooplankton, microcrustaceans, rotifers, worms and other aquatic feed animals
• Supplementary feed used in accordance with point 1.3 for penaeid shrimps and freshwater prawns (*Macrobrachium* spp.) may comprise a maximum level of 25% fishmeal and 10% fish oil derived from sustainable fisheries (Amendment Reg. 1358/2014)
• Conventional phytoplankton and zooplankton culture may be used as feed in larval rearing of organic juveniles (Amendment Reg. 1358/2014)
• Organic feed products of aquaculture origin
• Organic feed materials of animal origin
• Fishmeal and fish oil from organic aquaculture trimmings
1.2 Bivalve molluscs and other species which are not for human consumption but instead feed on natural plankton, the following rules shall apply:
• Such filter-feeding animals shall receive all their nutritional requirements from nature, except in the case of juveniles reared in hatcheries and nurseries
• The growing areas shall be either of high ecological status as defined by Directive 2000/60/EC or of good environmental status as defined by Directive 2008/56/EC or of equivalent quality to (a) the production zones classed as A in Regulation (EC) No 854/2004, until 13 December 2019, or (b) the corresponding classification areas set out in the implementing acts adopted by the commission in accordance with Article 18(8) of Regulation (EU) 2017/625, from 14 December 2019

(continued)

Table 8.1 (continued)

Acceptable Materials
1.3 Specific rules on feed for certain aquaculture animals
In the grow-out phase, fish in inland waters, penaeid shrimps and freshwater prawns and tropical freshwater fish shall be fed as follows
• (a) They shall be fed with feed naturally available in ponds and lakes
• (b) Where natural feed referred to in point (a) is not available in sufficient quantities, organic feed of plant origin, preferably grown on the farm itself, or algae may be used. Operators shall keep documentary evidence of the need to use additional feed
• (c) Where natural feed is supplemented in accordance with point, (b) the feed ration of Siamese catfish (*Pangasius* spp.) may consist of a maximum of 10% fishmeal or fish oil derived from sustainable fisheries
1.4 Milk and milk products
Produced under organic principles or nonorganic materials that have been authorised for use in organic production
• Raw milk
• Milk powder
• Skimmed milk, skimmed milk powder
• Buttermilk, buttermilk powder
• Whey, whey powder, whey powder low in sugar, whey protein powder (extracted by physical treatment)
• Casein powder
• Lactose powder
• Curd and sour milk
2. Feed materials of plant origin
Plant products shall not exceed 60% of total ingredients in the feed ration of carnivorous aquaculture animals
2.1 Aquatic origin
Produced under the organic principles
• Seaweed
• Multicellular marine algae or phytoplankton
• Microalgae
2.2 Land origin
Produced under the organic principles or nonorganic materials that have been authorised for use in organic production
2.1 Cereals, grains, their products and by-products:
• Oats as grains, flakes, middlings, hulls and bran
• Barley as grains, protein and middlings
• Rice germ expeller
• Millet as grains
• Rye as grains and middlings
• Sorghum as grains
• Wheat as grains, middlings, bran, gluten feed, gluten and germ

(continued)

Table 8.1 (continued)

Acceptable Materials
• Spelt as grains
• Triticale as grains
• Maize as grains, bran, middlings, germ expeller and gluten
• Malt culms
• Brewers' grains
2.2 Oil seeds, oil fruits, their products and by-products
• Rape seed, expeller and hulls
• Soya bean as bean, toasted, expeller and hulls
• Sunflower seed as seed and expeller
• Cotton as seed and seed expeller
• Linseed as seed and expeller
• Sesame seed as expeller
• Palm kernels as expeller
• Pumpkin seed as expeller
• Olives, olive pulp
• Vegetable oils (from physical extraction)
2.3 Legume seeds, their product and by-products
• Chickpeas as seeds, middlings and bran
• Ervil as seeds, middlings and bran
• Chickling vetch as seeds submitted to heat treatment, middlings and bran
• Peas as seeds, middlings and bran
• Broad beans as seeds, middlings and bran
• Horse beans as seeds middlings and bran
• Vetches as seeds, middlings and bran
• Lupin as seeds, middlings and bran
2.4 Tuber, roots, their products and by-products
• Sugar beet pulp
• Potato
• Sweet potato as tuber
• Potato pulp (by-product of the extraction of potato starch)
• Potato starch
• Potato protein
• Manioc
2.5 Other seeds and fruits, their products and by-products
• Carob
• Carob pods and meals thereof
• Pumpkins,
• Citrus pulp
• Apples, quinces, pears, peaches, figs, grapes and pulps thereof
• Chestnuts

(continued)

Table 8.1 (continued)

Acceptable Materials
• Walnut expeller
• Hazelnut expeller
• Cocoa husks and expeller
• Acorns
2.6 Forages and roughages
• Lucerne
• Lucerne meal
• Clover
• Clover meal
• Grass (obtained from forage plants)
• Grass meal
• Hay
• Silage
• Straw of cereals
• Root vegetables for foraging
2.7 Other plants, their products and by-products
• Molasses
• Seaweed meal (obtained by drying and crushing seaweed and washed to reduce iodine content)
• Powders and extracts of plants
• Plant protein extracts (solely provided to young animals)
• Spices
• Herbs
3. Feed materials of mineral origin
Of natural origin or nonorganic materials of mineral origin that have been authorised for use in organic production
3.1 Sodium
• Unrefined sea salt
• Coarse rock salt
• Sodium sulphate
• Sodium carbonate
• Sodium bicarbonate
• Sodium chloride
• Sodium formate
3.2 Potassium
• Potassium chloride
• Potassium iodine
3.3 Calcium
• Lithotamnion and maerl
• Shells of aquatic animals (including cuttlefish bones)

(continued)

Table 8.1 (continued)

Acceptable Materials
• Calcium carbonate
• Calcium lactate
• Calcium gluconate
• Calcium magnesium phosphate
3.4 Phosphorus
• Defluorinated dicalcium phosphate
• Defluorinated monocalcium phosphate
• Monosodium phosphate
• Calcium-magnesium phosphate
• Calcium-sodium phosphate
3.5 Magnesium
• Magnesium oxide (anhydrous magnesia)
• Magnesium sulphate
• Magnesium chloride
• Magnesium carbonate
• Magnesium phosphate
3.6 Sulphur
• Sodium sulphate
4. Feed additives
4.1 Nutritional additives
4.1.1 Vitamins
• Vitamins derived from raw materials occurring naturally in feeding stuffs
• Synthetic vitamins identical to natural vitamins for aquaculture animals
4.1.2 Amino acids
• Histidine (produced through fermentation) may supplement salmonid diets when the feed sources do not provide a sufficient amount of histidine to meet the dietary needs of the fish and prevent the formation of cataracts (Amendment Reg. 1358/2014)
4.1.3 Cholesterol
• Organic cholesterol may supplement shrimp diets where organic cholesterol is not available; nonorganic cholesterol derived from wool or shellfish or other sources may be used (Amendment Reg. 1358/2014)
4.1.4 Trace elements
• E1 iron: ferrous (II) carbonate, ferrous (II) sulphate monohydrate and/or heptahydrate ferric (III) oxide
• E2 iodine: calcium iodate, anhydrous calcium iodate, hexahydrate sodium iodide
• E3 cobalt: cobaltous (II) sulphate monohydrate and/or heptahydrate, basic cobaltous (II) carbonate, monohydrate
• E4 copper: copper (II) oxide, basic copper (II) carbonate, monohydrate copper (II) sulphate, pentahydrate
• E5 manganese: manganous (II) carbonate, manganous oxide and manganic oxide, manganous (II) sulphate, mono- and/or tetrahydrate

(continued)

Table 8.1 (continued)

Acceptable Materials
• E6 zinc: zinc carbonate, zinc oxide, zinc sulphate mono- and/or heptahydrate
• E7 molybdenum: ammonium molybdate, sodium molybdate
• E8 selenium: sodium selenate, sodium selenite
4.2 Zootechnical additives
• Enzymes and microorganisms
4.3 Technological additives
4.3.1 Preservatives
• E 200 sorbic acid
• E 236 formic acid (*)
• E 260 acetic acid (*)
• E 270 lactic acid (*)
• E 280 propionic acid (*)
• E 330 citric acid
(*) For silage: only when weather conditions do not allow for adequate fermentation
4.3.2 Antioxidant substances
• E 306 – Tocopherol-rich extracts of natural origin used as an antioxidant
• Natural antioxidant substances such as astaxanthin derived from organic crustacean shells may be used in the feeds for salmon and trout within the limit of their physiological needs
4.3.3 Binders and anticaking agents
• E 470 calcium stearate of natural origin
• E 551b colloidal silica
• E 551c kieselgur
• E 558 bentonite
• E 559 kaolinitic clays free from asbestos
• E 560 natural mixtures of stearites and chlorite
• E 561 vermiculite
• E 562 sepiolite
• E 599 perlite
• Conventional molasses
4.3.4 Silage additives
• Enzymes, yeasts and bacteria can be used as silage additives
• The use of lactic, formic, propionic and acetic acid in the production of silage is only permitted when weather conditions do not allow for adequate fermentation
• Emulsifying and stabilising agents
• Lecithin of organic sources
4.4 Certain substances used in animal nutrition
Substance listed must have been approved under Council Directive 82/471/EEC concerning certain products used in animal nutrition
• Organic yeast
• Yeasts: *Saccharomyces cerevisiae*, *Saccharomyces carlsbergensis* (fermentation by-products from microorganisms the cells which have been inactivated or killed)

(continued)

Table 8.1 (continued)

Acceptable Materials
• Phaffia yeast (as a source of astaxanthin – limitation not to exceed 10 ppm in fish) (not permitted for shrimp feed)
4.5 Substances for silage production
• Sea salt
• Coarse rock salt
• Whey
• Sugar
• Sugar beet pulp
• Cereal flour
• Molasses
Not acceptable
1. Feed materials of animal origin
• Fishmeal and fish oil from dedicated operations that are not independently certified as sustainable
• Fishmeal or other processed ingredients from the same taxa
• Meal and other processed ingredients from terrestrial animals
2. Feed materials of plant origin
• Nonorganic feed materials of plant origin, not listed here, that they have not been authorised for use in organic production
• Feeding stuffs that have been solvent extracted (except those extracted using ethanol and water)
• Genetically modified organisms or products and ingredients delivered from them
3. Artificial, synthetic or nature-identical pigments
4. Growth regulators, hormones and appetite stimulants
5. Synthetic binders

Adopted and modified from Mente et al. (2011); Council Regulation (EC) 2007, 2008, 2010, 2014; Regulation (EU) 2013

Table 8.2 A selection of similarities and differences to specific rules that apply to the feeding practices of aquatic animals reared in organic aquaculture as they have been defined in various international standards. For each standard/regulation, it is also reported the issue date and the country of origin

Similarities
Feed sources shall be based on the natural diet of the species to be certified and shall enable browsing and variety to mimic as much as feasibly possible the natural diet of the organisms being certified. **Australian Certified Organic Standard – ACOS (2016)**
Organic aquatic animals shall receive their nutritional needs from good quality, organic and other sustainable sources. Feeds for aquatic animals shall be formulated taking into account of the natural feeding habit, using organic ingredients, with appropriate ration size, to satisfy the nutritional requirements of the aquatic animal. **Hong Kong Organic Production, Aquaculture and Processing Standard (2015)**
Organic aquatic animals receive their nutritional needs from good-quality, organic sources, and aquatic animals shall be fed organic feed. **The IFOAM NORMS for Organic Production and Processing (2014) (International)**

(continued)

Table 8.2 (continued)

Organic aquaculture production provides a good quality diet balanced according to the nutritional needs of the organism. Feed is only offered to the organisms in a way that allows natural feeding behaviour, with minimum loss of feed to the environment. Feed is comprised of organically produced products, in situ nutrient sources, by-products from organic food processing and waste products from the fish industry. Aquaculture feeds shall contain 100% certified organic components or waste products only of aquatic origin. **NASAA Organic and Biodynamic Standard (2016) (Australia)**
Type, quantity and composition of feed must take into account the natural feeding methods of the concerned animal species. The activity level and the condition of the animals mainly give indications in this respect (e.g. corpulence factor, fat tissue). **Naturland Standard (2016) (Germany)**
The biological diversity of areas that are managed and adequate representation of naturally occurring organisms should be maintained. Operators should design feed rations to supply most of the nutritional needs of the animal from organic plants and animals appropriate for the digestive system and metabolism of the species. **OFDC Organic Certification Standards (2016) (China)**
Aquatic organisms should be provided with balanced dietary feeds to meet their nutritional needs. Feed stuffs shall come from materials not suitable for human consumption so that aquaculture production does not compete for human food. Feeding shall be performed in a way that respects the natural feeding behaviour of the stocks and mitigates the impact on the environment. **ACT Organic Standards (2016) (Thailand)**
Food programme: A food programme must be designed that meets the following requirements for organic aquaculture (a) that covers the nutritional needs of the animals in the different stages of their development. The plant material used in aquaculture feeds must be obtained from organic crops. ARTÍCULO 86.- **Ministry of Agro-Industry of Argentina, Resolution SENASA 374/2016**
Feeding and feed rations supplied to aquaculture animals shall be compatible with diets occurring in the natural environment and be designed according to the specific nutritional needs of each species. **Organic Aquaculture Standards (2012) (Canada)**
Organic aquaculture standards comply completely with the EU regulation for organic production. **KRAV Standards (2016) (Sweden)**
Organic aquaculture standards comply completely with the EU regulation for organic production. **Soil Association Standard (2016) (UK)**
Organic aquaculture standards comply with the IFOAM NORMS for Organic Production and Processing. **Organic Crop Improvement Association – OCIA (2013) (USA and Canada)**
Differences
For carnivorous animals, no more than 60% of the diet may comprise of plant products. Feed of agricultural origin shall be from sources produced and certified organic. Where such sources are not available, up to 5% of agricultural dry matter intake may be from nonorganic sources. Where marine food sources are used, a minimum of 50% of the total diet shall be comprised from by-products (not harvesting) of wild fish or marine organisms caught for human consumption. **Australian Certified Organic Standard – ACOS (2016)**
Where certified organic components or waste products are not available feed of conventional origin up to a maximum of 5% (by dry weight) including commercial fishmeal may be used. **NASAA Organic and Biodynamic Standard (2016) (Australia)**
In cases of scarcity or special conditions, according to the organic management plan, it will be allowed to use nonorganic food, a daily supply of 20%. **Interministerial Regulation n ° 28, of June 8, 2011. Minister of State for Agriculture, Livestock and Supply and the Minister of State for Fisheries and Aquaculture of Brazil**

(continued)

Table 8.2 (continued)

If certified organic ingredients are inadequate due to the early stage of local development of organic agriculture, raw materials of plant origin from conventional production or wild harvest may be used, but in any case, no more than 5% of the total dry weight of feeds for the entire year. The producer should use raw materials of plant origin as ingredients in aquaculture feed. But in case of carnivorous species, ingredients of plant origin shall not exceed 60% to prevent nutritional problems in the animals. **ACT Organic Standards (2016) (Thailand)**
If feed ingredients of animal origin (particularly fishmeal/oil) have to be used for the culture of carnivorous species with higher protein requirements, the following basic principles shall be respected: the animal components in feed shall, where acceptable for nutritional physiological reasons, be replaced by vegetable products. Where feed is used which is not produced in the course of the farm's aquatic food chains, the proportion of animal components in the feed shall be lower than 100%. Provisional maximum values are set in Part B. II. (Supplementary regulations for specific farming systems and animal species). **Naturland Standard (2016) (Germany)**
When the organic feed is not available, fishmeal and fish oil from by-products of conventional aquaculture or by-products of fish caught for human consumption may be used during a transitional period to be determined by SENASA. Such matters may not exceed (30%) of the daily ration. The feed ration may comprise up to a maximum of 60% of organic plant materials. **Ministry of Agro-Industry of Argentina, Resolution SENASA 374/2016**
When organic fishmeal or fish oil is not commercially available, it shall be preferentially sourced from trimming of fish already caught for human consumption in sustainable fisheries. Note: See Implementation of the 1995 FAO Code of Conduct for Responsible Fisheries. When nonorganic feed sources are used, they shall not exceed 80% of the action levels of the contaminants in feed. **Organic Aquaculture Standards (2012) (Canada)**
When organic feed is of inadequate quantity or quality, other feeds may be used under permission of HKORC-Cert and comply with the duration and conditions prescribed by HKORC-Cert and the requirements: (a) nonorganic aquatic animal protein and oil sources can only be used if the following conditions are satisfactorily implemented, 4.6.4.1, they are harvested from independently verified sustainable sources, and 4.6.4.2, they are verified to have contaminants below safety limits, and (b) animals may be fed with vitamins, trace elements and supplements from natural sources. **Hong Kong Organic Production, Aquaculture and Processing Standard (2015)**
Operators may feed a limited percentage of nonorganic feed under specific conditions for a limited time in the following cases: (a) organic feed is of inadequate quantity or quality; (b) areas where organic aquaculture is in early stages of development. **The IFOAM NORMS for Organic Production and Processing (2014) (International)**

Fishmeal Replacement

Salmonids and Marine Species (Sea Bream and Sea Bass)

It is well known that high-quality fishmeal provides a balanced amount of all essential amino acids, minerals, phospholipids and fatty acids reflected in the normal diet of fish (Hardy 2010; Lund et al. 2012) and hence ensures high utilisation by the fish and minimum discharge of nutrients to the environment. Hence, replacing fishmeal in diets for organic farming of salmonids and other marine fish species is not straightforward due to its unique characteristics including high protein content, excellent amino acid profile, high nutrient digestibility, high palatability, adequate amounts of micronutrients as well as general lack of anti-nutrients in fishmeal (Gatlin et al. 2007; Kaushik and Seiliez 2010; Krogdahl et al. 2010; Lund et al. 2012). Moreover, compared to salmonids, protein requirements of sea bass and sea bream are higher, reflecting their highly carnivorous nature (Oliva-Teles 2000; Peres and Oliva-Teles 2009).

Many studies have investigated the effects of replacing fishmeal with various plant protein ingredients (Carter and Hauler 2000; Altan and Korkut 2011; Borquez et al. 2011; Glencross et al. 2011; Lanari and D'Agaro 2005; Pereira and Oliva-Teles 2002; Pratoomyot et al. 2010; Sitjà-Bobadilla et al. 2005; Torstensen et al. 2008; Yang et al. 2011). Complete replacement by plant proteins has usually not been successful due to problems related to the anti-nutrient factors, altered patterns of amino acid uptake when replacing fishmeal with plant-based protein ingredients, lack of micronutrients and impairment of immunocompetence (Bendiksen et al. 2011; Borquez et al. 2011; Espe et al. 2006; Francis et al. 2001; Gatlin et al. 2007; Geay et al. 2011; Lanari and D'Agaro 2005; Larsen et al. 2012; Lund et al. 2011; Sitjà-Bobadilla et al. 2005). Concerning marine carnivorous species, studies on gilthead sea bream showed that 75% of protein could be provided by a large range of vegetable sources (corn gluten, wheat gluten extruded peas and rapeseed) without compromising digestive process, but the nutrient uptake significantly decreased up to this percentage (Santigosa et al. 2011b). A total substitution induced a strong reduction of the protease activity.

Experiments with plant proteins (soybean, rapeseed, corn gluten, wheat gluten, pea and lupin meals) have shown potential replacement of fishmeal with up to 25–35% (Negas and Alexis 1995; Carter and Hauler 2000; Pereira and Oliva-Teles 2003; Lanari and D'Agaro 2005; Hardy 2010; Enami 2011; Kaushik et al. 2004). In sea bream, it was observed that diets containing high levels (no more than 75%) of plant ingredients (corn gluten meal, wheat gluten, extruded peas, rapeseed meal and extruded whole wheat) did not affect fish growth performance and had only minor effects on quality traits of marketable fish (De Francesco et al. 2007). The organic feed ration may comprise a maximum of 60% of organic plant products, EC 889/2008. High replacement ratios require that anti-nutrients (such as trypsin inhibitors, tannins, lectins or glucosinolates) (Chebbaki et al. 2010) are efficiently removed from alternative plant protein ingredients to meet the high protein requirement of fish. The dietary content of indigestible substances should be minimised to optimise the efficiency of the feed. However, supplementation with synthetic amino acids is not allowed in feeds for organic aquaculture, EC No 834/2007. Furthermore, procedures for the removal of anti-nutrients have to follow organic rules. Finally, there is less availability of relevant organic plant sources to optimise the amino acid profile in comparison to conventional plant sources (Lund et al. 2011; Rembiałkowska 2007).

Extrusion processing could be used to obtain vegetable products with low levels of heat sensitive anti-nutritional factors and to increase the nutritional value of protein-containing ingredients (Chebbaki et al. 2010). Not all anti-nutrients are destroyed by heat treatment. Gossypol from cottonseed, glucosinolates in rapeseed meal and phytic acid in soybean, rapeseed and cottonseed meals are examples of harmful substances that need to be controlled by other means (Hardy and Barrows 2002). Furthermore, it is necessary to ensure that the dietary amino acid profile is optimised, for example, by adding free amino acids (only conventional feeds) and/or by combining available aquatic and plant protein sources with different amino acid composition (Francis et al. 2001; Kaushik and Seiliez 2010; Wilson 2002).

Nutrient requirements are most recently summarised in NRC (2011) "Nutrient Requirements of Fish and Shrimp". However, most of the data were obtained with juveniles under conditions regarded optimal including using high-quality experimental ingredients. Thus, there were no surpluses in the requirement data reported. A further safety margin is needed for nutrient loss in feed production, variation in content in feed ingredients and interactions between nutrients and ingredients. Further, research is needed to redefine nutrient requirements under suboptimum conditions such as due to climate change and environmental stressors or disease (Carter et al. 2010). Requirements also vary in different life stages of the fish. The requirements for amino acids, fatty acids, vitamins and minerals were determined with diets containing purified and chemically defined ingredients highly available to the fish. Nutrient bioavailability is variable in different feed ingredients and needs to be evaluated for every feed ingredient. Fish do not have absolute protein requirements but require the amino acids that compose the proteins. Atlantic salmon has documented requirements for the amino acids Arg, His, Ile, Leu, Lys, Met, Cys, Phe, Tyr, Thr, Try and Val, while Tau is not regarded as required (NRC 2011). Recent publications indicate higher requirements of lysine and threonine at the smolt stage (Grisdale-Helland et al. 2011, 2013) than the requirements reported as mean values by NRC. This may also be the case for other amino acids. The essential amino acid (EAA) requirements for optimal growth of Mediterranean fish species, such as sea bass, were Lys, Arg, Met, Cys, Try and Thr (Tibaldi and Kaushik 2005).

Dietary amino acid imbalance may be regarded as a primary cause of changes in feed consumption, and there is some evidence that voluntary feed intake in sea bass may be partially conditioned by limiting or excessive levels of certain diet EAA (Tibaldi and Kaushik 2005). Diets with lower proportions of tryptophan resulted in loss of appetite to juvenile sea bass, while diets lacking in methionine induced a reduction in feed intake (Thebault et al. 1985). Diets lacking in tryptophan could also be responsible of spinal deformities in sea bass fingerlings, as well as crystalline lens opacity and increased levels of Ca^{++} and Mg^{++} in the liver (Tibaldi and Kaushik 2005). Lysine and methionine are often the most limiting amino acids when fishmeal is replaced by plant protein sources (Mai et al. 2006). Farmed fish need a balanced dietary amino acid profile and especially the essential amino acids have to be provided in the diet in specific proportions (Carter and Houlihan 2001). If this is not the case, the surplus amino acids will be metabolised rather than used for protein synthesis and growth. A possible outcome is compromised fish welfare and environmental impact (due to increased nitrogen discharge) conflicting the organic principles, instead of being converted to fish meat. Therefore, a carefully balanced amino acid profile is important for the growth of the fish, as well as the minimisation of nitrogen discharge. Hence, production of well-balanced feeds for organic aquaculture is challenged by the limited options of only combining available aquatic and plant protein sources to balance the dietary AA profile.

The replacement of fishmeal by vegetable proteins is further complicated in finfish species because not only the overall dietary amino acid profile is important for efficient utilisation of amino acids but also the timing by which amino acids from different protein sources appear in the bloodstream after a meal (Larsen et al. 2012). Larsen et al. (2012) investigated differences in amino acid uptake, i.e. plasma free amino acid concentration patterns in juvenile rainbow trout (*Oncorhynchus mykiss*) fed either a fishmeal-based diet (FM) or a diet (VEG) where 59% of fishmeal protein (corresponding to 46% of total dietary protein) was replaced by a mixture of plant proteins from wheat, peas, field beans, sunflower and soybean. Results showed that the appearance of most amino acids (essential and non-essential) in the plasma was delayed in fish fed the VEG diet compared to those fed the FM diet. Essential and non-essential amino acids furthermore appeared more or less synchronously in the plasma in fish fed the FM diet, while the appearance was less synchronised in fish fed the VEG diet. Further there were 2.7 times more indigestible carbohydrates in the VEG diet than in the FM diet, which suggested that the uptake of amino acids was affected by dietary carbohydrates. In conclusion, the study showed that amino acid uptake patterns were affected when replacing fishmeal with plant-based protein ingredients.

A specific issue in replacing fishmeal by plant protein ingredients is that they contain significantly less phosphorus than fishmeal and the phosphorus is largely present as phytate phosphorus. Phytate phosphorus is not bioavailable to the fish as they lack the enzyme (phytase) necessary for releasing the phosphorus. Nonetheless, the use of phytase in fish feeds can help to reduce phosphorus waste (Lazzari and Baldisserotto 2008; Carter and Sajjadi 2011). Higher phytase levels in the feed were found to increase phosphorus, as well as nitrogen bioavailability and utilisation in plant-based diets used in sea bream aquaculture (Morales et al. 2013). However, the use of phytase in organic aquaculture could await authorisation of fermentation products produced by organic procedures for its permission in organic aquaculture. Phosphorous from fishmeal may have low bioavailability, and diets with theoretically adequate or surplus P levels, whatever vegetable and/or fishmeal based, may give P deficiency in salmon (Albrektsen et al. 2009).

According to EC 889/2008 fishmeal and fish oil from trimmings is prioritised as ingredient for feed for aquaculture animals, whether it is from organic aquaculture trimmings or ingredients of fish origin derived from trimmings of fish already caught for human consumption in sustainable fisheries. However, using fishmeal from trimmings in fish feed implies as well potential nutritional and environmental concerns. Fishmeal derived from trimmings might conflict with national environmental legislations due to too high P-concentrations. Fishmeal from trimmings is lower in protein and higher in phosphorus content compared with high-quality fishmeal (Eurofins; www.ffskagen.dk). The presence of carcass remnants (head, skin, bones) in trimmings also increases the phosphorus content of the fishmeal. Using this meal for feeding fish puts limitations on the inclusion level so as to comply with environmental legislation. F. ex. Danish environmental legislation only allows the phosphorus content of fish feed to be max. 0.9% (max. 1% on dry weight basis) (www.retsinformation.dk/Forms/R0710.aspx?id=140333). There are different chemical forms of phosphorus in the diet. Very significant differences were observed on the digestibility of the various forms (bone, phytin or organic phosphorus). Other factors, such as particle size and feed processing techniques, are also known to affect its digestibility (Azevedo et al. 1998).

Fishmeal and fish oil from organic aquaculture trimmings are not allowed in the feed for aquaculture animals of the same species. Hence, only limited quantities of trimmings from organic farming may be available and may only be sufficient for a very limited organic production, which may be below the critical level needed for sustainable manufacturing processes. The manufacturing process to obtain fishmeal and oil from trimmings is similar to that of wild-caught industrial fish (sand eel, blue whiting, etc.). However, due to the carcass remnants and the little remaining meat, the protein content of the meal from trimmings may be 67–70% and an ash content of about 15%. Further, the digestibility may be below 90% (pers. comm. Klaus Christoffersen, FF, Skagen, Denmark), while it should be at least 90% in a high-quality fishmeal. Carnivorous fish requires relative high dietary protein content, i.e. 38–48% of the diet, depending on fish size, with the highest requirement and quality for fry and broodstock. Indeed, the optimum protein level in the diets for sea bass juveniles was estimated to be around 50% (Hidalgo and Alliot 1988; Peres and Oliva-Teles 1999), independently from water temperature, while optimum protein level in the diet for gilthead sea bream fingerlings has been estimated to around 51% at 25 °C and 46% at 10–14 °C (Fountoulaki et al. 2005). This means that, to produce an adequate feed, the inclusion rate of fishmeal from trimmings should be high, which conflicts with the limitations of max. 0.9% dietary phosphorus content. Furthermore, the available organic plant sources are limited, and their amino acid profiles are not adequately balanced to make an optimum fish feed (Lund et al. 2011).

Crustaceans

Shrimps

The most important shrimp species in aquaculture are *Litopenaeus vannamei* (white shrimp) and *P. monodon* (giant tiger shrimp). Although they are all benthivore species, they have different diets in their natural habitats. Thus, *L. vannamei* is an omnivorous benthivore and mainly feeds on living preys and detritus (FAO 2011), and *P. monodon* is a carnivorous benthivore and mainly feeds on worms, crustaceans and molluscs (Tacon 2002; Piedad-Pascual 1984). These differences in feeding habits are due to the amount of enzymes in the digestive tract of the different shrimps. Carnivorous shrimps have proteolytic enzymes like trypsin and chymotrypsin, whereas herbivorous species have more glycolytic enzymes like amylase. This is why carnivorous shrimp have a greater ability to digest protein and herbivorous shrimp have greater ability to digest plant material.

Although it was estimated that optimum protein level for giant tiger shrimp was 40–50% of meal content (Conklin 2003; Mahmood et al. 2005), protein needs could change to sustain shrimp's maturation, reproduction and offspring quality (Wouters et al. 2001). Furthermore, the need for protein varies among species and the life stage of the animals. Juvenile stages have higher needs than older stages (subadults and adults), due to the different growth rate (Weir 1998). According to the available scientific literature, the needs for protein can vary for:

- *L. vannamei* between 20 and 30% of the dry matter in feed (Velasco et al. 2000; Cruz-Suárez et al. 2000; Kureshy and Davis 2000)
- *P. monodon* between 35% and 50% of the dry matter in feed (Fox et al. 1998; Cousin 1995; FAO 2011; Dayal et al. 2003; McVey 1993)
- *M. rosenbergii* between 30 and 38% of the dry matter in feed (Freuchtnicht et al. 1988; Reed and D'Abramo 1989)

Research on fishmeal substitution have also been conducted on carnivorous species of penaeid shrimps. Both partial and total substitutions of fishmeal with soybean meal (SBM) in the form of soy protein concentrate (SPC, 65% protein) were tested on giant tiger shrimp (*Penaeus monodon*) by Paripatananont et al. (2001). Generally, high dietary concentrations of soybean products in some species of shrimp negatively affected palatability. Anyway, Paripatananont et al. (2001) showed that up to 17.5% inclusion of SPC in shrimp feed did not adversely affect the feed intake and the growth rate, while further progressive levels of substitution lead to impaired body weight gain until the severe effects showed up at 100% substitution level. Other studies were conducted to substitute fishmeal and soybean meal (SBM) in shrimp aquaculture with microbial floc meal, produced in sequencing batch reactors (SBRs) (Kuhn et al. 2009; Emerenciano et al. 2012). Microbial biofloc has shown favourable nutritional quality and enhanced growth and production of shrimps (Kuhn et al. 2009; Emerenciano et al. 2012). Moreover, biofloc technology (BFT) showed to create economical and environment benefits via reduced water use, effluent discharges, artificial feed supply and improved biosecurity (Emerenciano et al. 2012). The EU production rule for organic aquaculture has set that their ration of supplementary organic feed may comprise a maximum of 25% of fishmeal and 10% of fish oil derived from sustainable fisheries (EU 2009, 2013, 2014).

Alternative Feed Ingredients Sources for Organic Aquafeeds

In addition to the plant proteins, there are several other potential feed ingredients, such as microbial organisms (bacteria, fungi, microalgae), terrestrial animal by-products (PAP, blood meal), wild-harvested and/or cultured annelid worms, insect larvae/pupae and gastropods (e.g. golden apple snail) which also may be candidates to replace fishmeal in aquaculture feed in the future (Perera et al. 1995; Bergleiter et al. 2009; Sørensen et al. 2011; Van Huis and Oonincx 2017).

Microbial ingredients, i.e. products from bacteria, yeast and microalgae, are expected to have an important potential in future salmonid and other carnivorous fish species (such as sea bass and sea bream) feed. A special aspect of some of these products is that they can be produced with different kinds of waste as raw material and thus contribute to circular economy by recycling of valuable nutrients. A large number of products, produced from various single cell organisms grown on different materials, have been investigated (e.g. Anupama and Ravindra 2000; El-Nawwi and El-Kader 1996; Mathews et al. 2011; Rajoka et al. 2006). Depending on type of organism, the proximate composition and amino acid profile can be much similar to that of fishmeal (Øverland et al. 2010). A number of products have been tested as protein sources in fish feeds, and the suitability varies among the different products, inclusion levels and fish species tested (Oliva-Teles and Gonçalves 2001; Li and Gatlin 2003; Berge et al. 2005; Aas et al. 2006; Palmegiano et al. 2009a, b; Romarheim et al. 2011; Øverland et al. 2013).

Microalgae as raw matter or a feed ingredient for fish have also gained interest, as they are the natural start of the food chain in the oceans. Microalgae are fed on by zooplankton, which again is fed on by fish. The idea is to harvest from the first trophic level or cultivate in closed systems and provide feed ingredients for farmed fish by culturing microalgae. Although the benefit of different microalgae in an organic feed use should be demonstrated scientifically for each aquaculture species, a mass production and an economic model should be developed. Living microalgae are used in aquaculture for fish feeding during the early stages, and the benefit of addition of marine *Isochrysis* sp. to cultivated zooplankton for European sea bass was demonstrated on immune and digestive system (Cahu et al. 1998). Modern process and algae cultivation in photobioreactors or fermentation systems can provide algae under a flour form, which can be used with the same form as fishmeal to produce formulated pellets. The chemical composition of microalgae varies depending on species, cultivation parameters and the potential as a feed ingredient varies accordingly (Skrede et al. 2011). Diets containing the microalga T-Iso (*Isochrysis* spp.) resulted in improved growth of gilthead sea bream juveniles, and the chemical composition of sea bream fillets also met the needs of consumers, although the level of proteins was lower compared to a conventional diet and the level of fats was higher. T-Iso resulted in high digestibility and supported the best performance of fish fed on a diet based on 70% of microalgae, probably due to its high protein efficiency (Palmegiano et al. 2009a, b).

Algal addition in diets have been conducted with rainbow trout fry (*Oncorhynchus mykiss*) using a biomass of photosynthetic microorganisms composed by a mixture of *Scenedesmus* sp. and *Chlamydomonas* (29.6% of crude protein) from a fish farm sedimentation pond. The results showed that a maximum of 12.5% of algal biomass could be incorporated in the feed for rainbow trout fry (*O. mykiss*) without negative effects on growth and body content of lipids and energy of fish (Dallaire et al. 2007). The evaluation of the microalgae *Isochrysis* sp. in partial substitution of fishmeal in gilthead sea bream (*Sparus aurata*) pellets showed better performances than control diets. The best performances of fish fed on 70% algae diet were probably due to the protein composition and the amino acid profile in comparison to other diets

(Palmegiano et al. 2009a, b). Other algae species as *Tetraselmis suecica* could replace up to 20% of European sea bass protein without hampering growth performance and major quality traits fish (Tulli et al. 2012).

Processed animal protein (PAP) is an important ingredient in feeds and provides a valuable source of animal by-product utilisation (Karapanagiotidis et al. 2018). Nutritional quality of rendered animal protein ingredients is affected by composition, freshness of raw materials and processing conditions. PAP has a high nutritional value making it an excellent alternative to imported proteins such as soya. It has a significantly higher protein value (45–90% on a fed basis) than plant feed ingredients. PAP contains 10% phosphorus, which is low in relation to the content of amino acids. Blood meal is also a feed ingredient with high protein content (80% in full blood) and excellent protein digestibility (Bureau et al. 1999). It has high content of lysine and histidine, while the content of isoleucine is low (El-Haroun and Bureau 2007; Breck et al. 2003). While there may be consumer and producer concerns about the feeding of PAP to fish, due to the potential transmission of prions, the scientific panel opinion published by the European Food Safety Authority (EFSA) in 2011 concluded that processed animal protein in feed for food producing non-ruminants, respecting the proposed ban on intraspecies recycling, presents a negligible risk to human health (EFSA 2011). In addition, Commission Regulation (EU) No. 56/2013 of 16 January 2013 amending Annexes I and IV to Regulation (EC) No. 999/2001 of the European Parliament and of the Council laid down rules for the prevention, control and eradication of certain transmissible spongiform encephalopathies.

The use of insects as a source of protein in fish and crustaceans diets is also being explored (Karapanagiotidis et al. 2014; Henry et al. 2015; Cummins et al. 2017; Magalhães et al. 2017; Devic et al. 2018; Piccolo et al. 2017; Iaconisi et al. 2017; Gasco et al. 2018; Rumbos et al. 2018). The chemical composition of prepupae larvae varies with species, age, method of processing and the substrate the maggot is produced on (St-Hilaire et al. 2007a, b; Aniebo and Owen 2010; Sealey et al. 2011). The nutritional value of insects as feeds for fish, poultry and pigs has been recognised for some time in China, where studies have demonstrated that insect-based diets are cheaper alternatives to those based on fishmeal. The insects used are the pupae of silkworms (*Bombyx mori*), the larvae and pupae of houseflies (*Musca domestica*) and the larvae of the mealworm beetle, *Tenebrio molitor*. Feeding experiments with *T. molitor* have shown that insect meal obtained from this species can be included at various percentages without negative effects on growth performance in the diet of the European sea bass, *Dicentrarchus labrax* L. (Gasco et al. 2016), and the gilthead sea bream, *Sparus aurata* L. (Piccolo et al. 2017). Silkworm pupae are an important component of cultured carp diets in Japan and China. Dried ground soldier fly larvae have been fed to chickens and pigs with no detrimental effects (Newton et al. 1977; Hale 1973). In recent years there has been some interest in the use of housefly maggot meal as a substitute for fishmeal in tilapia and African catfish diets (Adesulu and Mustapha 2000; Fasakin et al. 2003; Ajani et al. 2004; Ogunji et al. 2006). Bondari and Sheppard (1987) observed that channel catfish and blue tilapia fed on soldier fly larvae for 10 weeks were acceptable as food by consumers. Sealey et al. (2011) indicated that black soldier fly prepupae reared on dairy cattle manure, and trout offal can be used to replace up to 50% of fishmeal in the diet of rainbow trout for 8 weeks without significantly affecting fish growth or the sensory quality of rainbow trout fillets. Comparison of amino acid profiles in the test diets that were replacing fishmeal with black soldier fly larvae for the Pacific white shrimp *L. vannamei* presented limiting amino acids in black soldier fly larvae meal suggesting future strategies for increasing its dietary fishmeal substitution (Cummins et al. 2017). Growth and organoleptic quality was not affected when common carp were fed on non-defatted silkworm pupae, a major by-product of the sericulture industry in India (Nandeesha et al. 2000). Ng et al. (2001) demonstrated that *T. molitor* larvae meal was highly palatable to the African catfish (*Clarias gariepinus*) and could replace up to 40% of the fishmeal component without reducing growth performance.

St-Hilaire et al. (2007a, b) described a study in which they determined if black soldier fly (*Hermetia illucens*) prepupae and housefly pupae could be used as a partial replacement for fishmeal and fish oil in rainbow trout (*Oncorhynchus mykiss*) diets. Their data suggested that a rainbow trout diet in which black soldier fly prepupae or housefly pupae constituted 15% of the total protein had no adverse effect on feed conversion efficiency over a 9-week feeding period. However, rainbow trout fed on black soldier fly diets low in fish oil had reduced levels of n-3 fatty acids in the muscle. According to the researchers, modifying the diet of the fly larvae could improve digestibility and fatty acid content of the prepupae, which in turn could enhance the fatty acid profile of the fish fed on the fly prepupae. The use of the black soldier fly in manure management yields abundant numbers of fly prepupae. The authors of the study suggested that fly prepupae may be an economical and sustainable feed ingredient for carnivorous fish diets. However, before fly prepupae can be used commercially in rainbow trout diets, a larger trial over a longer period should be conducted to confirm their preliminary results. Yellow mealworm meal could partially (35%) replace fishmeal in the diet of European sea bass (*Dicentrarchus labrax*) without affecting mortality or growth, but replacing 70% of the fishmeal did depress growth (Gasco et al. 2016).

Fish Oil Replacement

Salmonids and Marine Fish (Sea Bream, Sea Bass)

Fish oil is a major natural source of the long-chain n-3 HUFAs eicosapentaenoic acid (EPA) and docosahexaenoic acid (DHA), which can be synthesised by salmonids and by other carnivorous marine species only at a limited rate and thus are required in the diet. Omega-3 (n-3) HUFAs are produced by marine phyto- and zooplankton, which are consumed by the wild marine fish larvae (Baron et al. 2013). Hence, fishmeal and fish oil are “strategic ingredients” to be used at critical stages of the lifecycle, when optimum performance is required. Furthermore, farmed larvae and broodstock need the "strategic ingredients" to secure optimum development, growth and reproduction.

Atlantic salmon and rainbow trout can convert ALA to EPA and DHA, but the conversion is not very efficient for marine stages (Ruyter and Thomassen 1999, 2000a, b, c; Tocher et al. 2000; Bell et al. 2001; Bell and Dick 2004). Therefore, their essential fatty acid (FA) requirements must be provided by dietary EPA and DHA to obtain good growth and health (Ruyter et al. 2000a, b, c). It is also shown that the ability to convert 18:3 n-3 to EPA and DHA may be induced by plant oil inclusion in fish diets (Moya-Falcon et al. 2005) and that the conversion is higher in the freshwater stage prior to smoltification, than at later post smolt life stages in seawater (reviewed by Bell and Koppe 2011). Bell and Dick (2004) showed that DHA synthesis was at its highest in rainbow trout in the period immediately after start feeding and then declined over a period of a few weeks. It has also been shown that the capacity for conversion of ALA to EPA and DHA and the gene expression of the Δ5- and Δ6-desaturase activities in salmon were depressed when fed high dietary levels of fish oil (FO), while vegetable oils to a certain degree increased the capacities (Ruyter et al. 2003; Moya-Falcon et al. 2005; Kjær et al. 2008). Several studies have shown that there are species differences in the capacities for the conversion of essential FAs (Sargent et al. 2002). A recent study (Berge, personal communication) showed that rainbow trout had apparent retention of 22:6 n-3 in the range 129–194%, while the salmon obtained values in the range 86–120%. It is well known that rainbow trout can synthesise DHA from ALA (Buzzi et al. 1996; Bell et al. 2001; Bell and Dick 2004), and this process is also taking place when fish are fed with diets high in 22:6 n-3 (Buzzi et al. 1996). The studies indicated that rainbow trout had better ability than Atlantic salmon to convert the shorter vegetable n-3 FAs to the important long-chain EPA and DHA.

Diets based on vegetable oils have generally shown good growth results in salmon but with major challenges in the body lipid composition (Thomassen and Røsjø 1989; Sargent et al. 2002; Grisdale-Helland et al. 2002; Torstensen et al. 2005). Altered FA composition may affect the fish health. Omega-3 (n-3) fatty acids have important biological functions in the fish (Montero et al. 2010; Torstensen et al. 2013), and a change in dietary fatty acid composition is expected to affect fish performance and health. The n-3 fatty acids serve as the building blocks of cell membranes, regulate gene expression and are precursors of a range of bioactive substances that regulate inflammation, physiology and satiation. By optimising dietary fatty acid composition, the retention of EPA+DHA can be optimised and thereby improving fish health as well as securing the farmed salmon as a good source of EPA+DHA for human consumption. The balance between n-6 and n-3 FAs in the diet seems to be important, as the pro-inflammatory eicosanoids from the n-6 family are more abundant and have greater biopotency than their n-3 homologues (Lands 1992). This may have impact on several aspects of fish health. 18:2 n-6 and 18:3 n-3 also compete for the same enzyme systems, for synthesis of long-chain PUFAs. However, the optimal n6/n3 ratio for conversion of 18:3 n-3 to EPA and DHA in salmonids is not known. The feed oils for the future may consist of a mix of different oils that provide an optimal dietary ratio between groups of FAs (C18 n-6/C18 n-3/C20 n-3/C22 n-3) and utilises the innate ability of the fish to produce EPA and DHA. The optimal mix will be defined by the trade-off between high rate of deposition (retention) of dietary long-chain n-3 HUFAs and high absolute level of these fatty acids, which are to some extent conflicting aspects. The optimal oil mix (FA combination) may vary for different life stages and in different environments.

Salmonids prior to smoltification have higher Δ5- and Δ6-desaturase activities than fish in seawater (Sargent et al. 2002). Oils that contain appropriate levels of C18 n-3 fatty acids, but relatively low content of C18:2 n-6, are preferred. Linseed oil, as well as less commonly used oils like camelina oil and chia oil, are candidates that are rich in C18 n-3. However, one must be aware that it is shown that high dietary levels of ALA seemed to inhibit its own conversion to DHA in Atlantic salmon, and the dietary level must therefore be optimised (Ruyter et al. 2000a, b). Camelina oil has been tested in diets for Atlantic salmon (Hixon et al. 2014), and 100% exchange of fish oil with camelina oil caused a small but not significant drop in growth rate. Salmon lipid composition reflected the dietary fatty acid profile, with a higher content of 18:3 n-3 in fish fed the camelina oil diet. Echium (*Echium plantagineum*) oil is another promising oil, with high content of C18:4 n-3, one step further to a long-chain n-3 FA compared to 18:3 n-3. Echium oil has been tested in diets for Atlantic salmon (Miller et al. 2008; Codabaccus et al. 2011); there was higher biosynthesis of eicosatetraenoic (20:4 n-3) and 20:5 n-3 by salmon parr and smolts in the Echium oil group compared to FO and canola oil groups.

For the replacement of fish oil, marine fish species, such as sea bass and sea bream (Geay et al. 2011), have lower tolerance to vegetable oil compared to freshwater or anadromous fish species such as salmonids. This lower adaptation of marine fish species to vegetable oil can be linked to their lower efficiency in synthesising LC-PUFA from n-3 to n-6 precursors present in plants (Geay et al. 2011). A high or total substitution of fish oil by plant oils induced decreases in growth rate of gilthead sea bream and European sea bass (Geay et al. 2010; Montero et al. 2010). Studies on sea bream showed that a replacement up to 66% of fish oil can be operated by a vegetable source with comparable results. Nutrient absorption in fish intestine was negatively modified for a total substitution of fish oil by vegetable oil. An impaired digestion was observed induced by an accumulation of lipidic droplet in the fish intestine (Santigosa et al. 2011a) probably due to a saturation of fish assimilation sites. Indeed, LC-PUFA, used as structural components of cell membranes, are also the principal precursors of eicosanoids that are involved in many physiological processes such as osmoregulation, immune responses, blood coagulation and reproduction (Bell et al. 1997; Geay et al. 2011). The lower nutritional value in the flesh of marine fish fed vegetable diet is generally due to the low content in EPA end DHA.

Isochrysis sp., partially substituted in sea bream diets, showed to be a good source of polyunsaturated fatty acids and in particular of docosahexaenoic acid (DHA) (Palmegiano et al. 2009a, b).

Progressive substitutions of fish oil with cottonseed oil (CSO) did not affect fish growth, feed conversion ratio and protein utilisation, but hepatosomatic and visceral fat indexes increased with increasing dietary CSO (Eroldogan et al. 2012). CSO, being a rich source of n-6 PUFA, may affect hepatocyte vacuolation and lipid infiltration, and this could be likely ascribable to the reported lipogenic effect of 18:2 n-6, as suggested by Montero and Izquierdo (2010).

Partial substitution (50%) of fish oil, respectively, with sesame oil (SO), canola oil (CO) and soybean oil (SBO) in sea bass did not influence the whole body fatty acid composition in terms of saturated fatty acids (SFA), polyunsaturated fatty acids (PUFA), eicosapentaenoic acid (EPA, 20:5n-3) and docosahexaenoic acid (DHA, 22:6n-3) contents (Özşahinoğlu et al. 2013). The diet that showed the best growth performances was the one using sesame oil as substitute. Also, the partial substitution (50%) of fish oil with soybean oil in both sea bass and rainbow trout had no effects on either hepatic lipid droplets accumulation or the degree and pattern of vacuolisation (Figueiredo-Silva et al. 2004).

Higher levels of fish oil substitution were applied in both sea bass and sea bream by Izquierdo et al. (2003) using, respectively, soybean oil (SO), rapeseed oil (RO) and linseed oil (LO) or a mixture (Mix) of them. Feed intake was not influenced by the vegetable oils as well as fish growth. Fatty acid composition of the liver and muscle reflected that of each single diet, but utilisation of dietary lipids differed between these two tissues and was also different for the different fatty acids. Sea bass liver showed much higher lipid contents than seabream, due to a greater accumulation of saturated (mainly 16:0) and monounsaturated (mainly 18:1n-9) FAs. Muscle lipid contents were very similar for both species.

Mourente et al. (2005) showed that vegetable oils such as rapeseed, linseed and olive oil can potentially be used as partial substitutes for dietary FO in European sea bass culture, during the grow out phase, without compromising growth rates, but may alter some immune parameters. Indeed, an alteration of the non-specific immune function was observed, and the number of circulating leucocytes was significantly affected, as well as the macrophage respiratory burst activity. Accumulation of large amounts of lipid droplets were observed within the hepatocytes in relation to decreased levels of dietary n-3 HUFA, although no signs of cellular necrosis were evident. Inclusion of vegetable oils (rapeseed, linseed and olive oil), up to 600 g kg^{-1} of dietary oil, significantly reduced EPA and DHA and increased linoleic and linolenic presence in sea bass flesh. The time required to restore individual fatty acids to values like those in fish fed fish oil were different for each fatty acid. In the same study, Mourente et al. (2005) also observed that some fatty acids are selectively retained or utilised. There was a selective deposition and retention of DHA because flesh DHA concentrations were always higher than diet concentrations, as observed also for salmonids. Linolenic (LNA; 18:3 n-3), linoleic (LA; 18:2 n-6) and oleic (OA; 18:1 n-9) acids concentration significantly increased in flesh lipids following the fish oil substitution with vegetable oils. This should be considered, as reducing n-6 PUFA, largely as linoleic acid, has benefit implication in the human diets.

Microalgae and especially marine species are promising alternative fatty acid sources of interest in aquaculture feed. It was shown that microalgae oils from *Isochrysis*, *Nannochloropsis*, *Phaeodactylum*, *Pavlova* and *Thalassiosira* contain sufficient n-3 LC-PUFA to serve as an alternative for fish oil (Ryckebosch et al. 2014). However, culturing microalgae, by fermentation like yeasts under controlled conditions that give a higher production efficiency for production of n-3 LC-PUFA, is still regarded as being in the development stage resulting in low production volumes and higher costs which, at present, are far more expensive than for fish oil and meal (Sprague et al. 2016). Evaluation of microalga *Isochrysis* sp. in partial substitution of fishmeal revealed a positive effect on gilthead sea bream (*Sparus aurata*) performances and chemical composition of fillets. Best fish performances were observed when fish fed on 70% algae diet probably due to highest amount of saturated fatty acids, mainly due to myristate and palmitate acid (Palmegiano et al. 2009a, b). The use of heterotrophic algae source *Schizochytrium* or *Crypthecodinium cohnii* in the early sea bream stage showed an important potential of these strains as alternative DHA sources for fish feed in microdiets and also pointed out the necessity of EPA sources to completely replace fisheries-derived oils (Atalah et al. 2007; Ganuza et al. 2008). Further alternative marine sources of n-3 LC-PUFA include the use of krill or calanoid copepods, although again production volumes are expected to be low and there is concern over the possible effects that harvesting down the trophic chain may have on higher trophic species that are normally reliant on these zooplankton (Sprague et al. 2016).

Crustaceans

Shrimps

Lipids are also essential components of the diets for shrimps and are mainly used for direct energy production and cell membrane building. Typically, crustaceans have limited ability to ex novo synthesise HUFA, as observed in marine fish (Mourente 1996), at least at the beginning of maturation. Similarly, there are difficulties for the ex novo synthesis of cholesterol (Kanazawa et al. 1985), useful to synthesise steroid hormones (Kontara et al. 1997). For *L. vannamei* and *P. monodon*, the optimal lipid level was between 6 and 8% of the feed dry matter (Alday-Sanz 2011; Tiwari and Sahu 1999) but should not be above 10% (Glencross et al. 2002) or below 2% (Chen 1998). Some lipids are more important than others because they cannot be synthesised de novo or not in sufficient amounts by shrimps. Phospholipids (e.g. lecithin) and cholesterol are the two main categories of essential lipids for shrimps. They are also used as emulsifiers for lipid digestion. Without phospholipids in their diet, shrimps are unable to digest lipids properly.

According to the available scientific literature, the need for phospholipids for *P. monodon* is 1% of the diet for post-larvae (Paibulkichakul et al. 1998) and 1.25% for juveniles (Chen 1993); and for *L. vannamei* the requirements for lecithin and cholesterol are linked together. Cholesterol is a ring compound, which is part of cell membranes and is also necessary in the moulting process. According to the literature, the need for cholesterol varies among the different species of shrimps and according to the different life stages. For *P. monodon*, cholesterol need is lower, but it is crucial and cannot be replaced. Requirements were 1% of the diet for post-larvae (Paibulkichakul et al. 1998) and 0.17% of the diet for juveniles (Smith et al. 2001).

Though, for *M. rosenbergii*, cholesterol needs were high, 0.3–0.6% of the diet (Sahu 2004), but this species is able to use phytosterols contained in plant instead of cholesterol as ecdysone precursors, so the amount added in the diet can be reduced significantly (Mitra et al. 2005).

For *L. vannamei*, there is a relationship between cholesterol and phospholipids. A diet with no phospholipids required 0.35% cholesterol, whereas a diet with 5% phospholipids required only 0.05% cholesterol (Gong et al. 2000). A good combination seemed to be 0.15% of cholesterol for 1% or more phospholipids.

It is important to keep focus on human health related to eating (organic) aquaculture products, including high content of long-chain n-3 fatty acids (EPA and DHA) currently sourced from fish oil. However, there seems to be a challenging conflict between organic principles and the health aspects of n-3 fatty acids in fish (human health), as well as the performance of fish and shrimps, and the regulation requesting exchange of fish oil with plant oil mostly low in n-3 fatty acids, which compromises the promotion of (organic) fish as healthy. An option might be development of plant oils rich in n-3 fatty acids. However, non-GMO plants are not able to produce n-3 fatty acids longer than 18C and cannot supply EPA and DHA.

Minerals

Salmonids and Marine Species (Sea Bream, Sea Bass)

Information on requirements of minerals and vitamins for salmonids is limited, but these species are still among the best documented in comparison to marine finfish such as sea bass and sea bream.

Minerals are divided into macrominerals (P, Ca, Mg, K, Na, Cl; required in relatively large amounts) and micro-minerals (Zn, Cu, Se, Mn, Mb, Fe, I, Cr, Co). Phosphorous and calcium are needed in large amounts for skeletal tissue, as well as other functions. In Atlantic salmon, skeletal deformities are seen occasionally because of mineral (P) deficiency. There are few available data on the mineral requirements of marine finfish such as sea bass and sea bream. For sea bream Oliva-Teles (2000) reported a dietary phosphorus requirement around 0.75%. Bioavailability of phosphorus is highly variable among feedstuffs and is higher in animal than in plant feedstuffs (Oliva-Teles 2000). This is due to the major proportion of phosphorus in plants being stored as phytate, which is not available to animals. Gomes da Silva and Oliva-Teles (1998) estimated the apparent digestibility coefficients (ADC) of phosphorus for sea bass juveniles: the ADC of phosphorus of animal feedstuffs averaged 81% while that of soybean was only 38%. Substituting fishmeal with plant meal can markedly affect the mineral composition of feeds and may require additional mineral supplementation. Dietary minerals supplemented in the organic form in diets for sea bream could be reasonably considered more effective than the inorganic and encapsulated forms of supply (Domínguez et al. 2017).

Dietary supply in organic production is preferably from natural origin, but chemically well-defined analogic substances may be authorised for use if the natural substances are unavailable (EC 834/2007). There are factors that complicate the assessment of dietary requirement of minerals and vitamins. Fish may absorb some of these nutrients from the water, and nutrients may leach from diet to water, difficulties in producing good test diets, and lack of knowledge on bioavailability of the nutrients. The current practice is to add nutrients to the diet, based on existing knowledge, but with a significant safety margin. Because of the weak evidence, dietary requirements may be underestimated for some of these nutrients.

Crustaceans

Shrimps

Calcium and phosphorus are two of the major constituents of the inorganic portion of feed for shrimps (Davis et al. 1993). Shrimp can absorb calcium from the water via drinking or absorption from the gills, epidermis or both. On the other hand, phosphorus concentration in natural water is generally too low, making the dietary phosphorus income essential for shrimps. Davis et al. (1993) observed in *L. vannamei* that in absence of dietary calcium supplementation, the adequate dietary phosphorus amount was 0.34%, although the minimum level of dietary phosphorus for maximum growth of *L. vannamei* is dependent on the calcium content in the diet. In addition, shrimp diet containing 3% or more of calcium should be avoided. In *P. monodon* Ambasankar et al. (2006) estimated that the best zoo technical performances were recorded by the diets supplemented with 1.0 and 1.5% phosphorus.

Vitamins

Salmonids and Marine Species (Sea Bream, Sea Bass)

Vitamins are needed in trace amounts in the diet to maintain normal growth, reproduction and health. Characteristic deficiency signs are seen in mammals in the absence of vitamins, but in fish the deficiency signs are less specific. The requirements are affected by size, age, growth rate, environmental factors and nutrient interactions. Of the water-soluble vitamins, the B-complex is needed in relatively small amounts, while choline, inositol and vitamin C are needed in larger amounts. Status on vitamin requirement knowledge is reviewed by NRC (2011) for salmonids and marine species. Various marine microalgae strains might provide excess or adequate levels of the vitamins for aquaculture food chains (Brown et al. 1999; Coutinho et al. 2006).

Available data on vitamin requirements of marine finfish such as sea bass and sea bream are very scarce. A dietary requirement for vitamin B6, pyridoxine, has been demonstrated in several species of freshwater and marine fish. Kissil et al. (1981) reported that signs of pyridoxine deficiency were manifested in sea bream as growth retardation, high mortality, poor feed conversion, hyperirritability coupled with erratic swimming behaviour and degenerative changes in peripheral nerves. The authors also estimated the dietary pyridoxine level at and above which no deficiency signs appeared: 1.97 mg/kg dry diet. Bioavailability of ascorbic acid (AA) esters, such as the phosphate forms, has been found to be high in several fish. The minimum dietary ascorbic acid stable phosphate forms requirement reported in literature was in the range of 10–20 mg of AA/kg for freshwater fish and 12.6–47 mg of AA/kg for some marine fish (as reviewed by Fournier et al. (2000)). Ascorbic acid is necessary for the hydroxylation of proline, leading to hydroxyproline (HyPro), which is involved in collagen synthesis. Except for some Cyprinids and some Acipenserids, most finfish cannot synthesise AA. Documented pathological effects of vitamin C (ascorbic acid) in sea bream are reported by Alexis et al. (1997). Such pathological signs appeared in all fish fed the vitamin C-deficient diet, extensive tubular damage, glomerulonephritis, and inflammatory response of the haemopoietic tissue producing granuloma, while the gross deficiency signs observed were anorexia, scale loss, depigmentation and internal and external haemorrhages. Henrique et al. (1998) estimated that the ascorbic acid requirements for sea bream were less than 25 mg/kg. While for juvenile European sea bass, the minimum dietary AA requirement reported by Fournier et al. (2000) to maintain normal skin collagen concentration and maximal growth was 5 mg of AA/kg, apparently below the requirement of other fish, although higher levels were required based on whole body hydroxyprolin and liver ascorbic acid concentration. Kaushik et al. (1998) tested the recommendations for salmonids of NRC (1993) for vitamin requirements in sea bass. The authors confirmed the applicability of the NRC salmonids recommendations in diets for sea bass, although in semi-purified diets a slightly higher supply was necessary to allow satisfactory growth rates. Among natural antioxidants, vitamin E has been found to offer a protective role against the adverse effects of reactive oxygen and other free radicals. Gatta et al. (2000) demonstrated that a level of 942 mg kg^{-1} in the diet was sufficient for sea bass.

Crustaceans

Shrimps

Vitamins have pivotal roles to ensure good survival rates in aquaculture also in shrimp's dietary. Indeed, vitamin C-deficient diets in *L. vannamei* caused reduced survival rates, while growth was not affected (He and Lawrence 1993a). Moreover, it was observed that whole-body ascorbic acid content in shrimp increased as dietary vitamin C increased. He and Lawrence (1993a) estimated also that the minimum dietary vitamin C levels required for normal survival of *L. vannamei* specimens of 0.1 g and 0.5 g were, respectively, 120 mg ascorbic acid equivalent (AAE)/kg and 90 mg AAE/kg, showing that dietary vitamin C requirement of *L. vannamei* decreased with increased size. Furthermore, there are evidences that dietary ascorbate enhances immune responses in *L. vannamei* (Lee and Shiau 2002). Vitamin E, as a fat-soluble compound, is the most effective lipid-soluble antioxidant in biological membranes, where it contributes to membrane stability (He and Lawrence 1993b). Moreover, it protects cellular structures against oxidative damages from oxygen-free radicals and reactive products of lipid peroxidation. Lee and Shiau (2004) demonstrated that a level of 85–89 mg kg^{-1} of vitamin E was required for maximal growth and non-specific immune responses of *P. monodon,* while 179 mg kg^{-1} of vitamin E was required to maximise tissue vitamin E concentration.

Pigments: Astaxanthin

Salmonids and Shrimps

Astaxanthin is the preferred carotenoid for pigmentation in salmonids causing the red colour of the muscle and is found naturally in potential feed ingredients like shrimp, krill, calanus, capelin oil and some yeasts and algae. It is also known as a potent antioxidant. The carotenoids are mobilised from muscle to skin and ovaries in maturing fish, but the role in reproduction is not fully understood. According to Torrissen and Christiansen (1995), dietary carotenoids are required in fish diets, suggestively with a metabolic role like that of vitamin E and A. Astaxanthin from microalgae, mainly extract by a green microalgae *Haematococcus pluvialis*, attracts considerable attention for its biological properties such as the antioxidant activity, colouring agent and lipid sources for farmed fish feed (Choubert et al. 2006; Fujii et al. 2006). Main supply for salmonid culture is synthetic astaxanthin. An eco-efficiency study performed by Gensch et al. and published by BASF around 2004 indicated that the sustainability in production of astaxanthin for pigmentation of salmon was best in synthetic production and poorer in yeast and algae products. The factors considered were surface use, energy use, emissions, raw material use, risk potential and toxicity potential. The pigmentation of shrimps, as well as for salmonids, is influenced by astaxanthin dietary intake. Indeed, an optimal pigmentation in *P. monodon* is guaranteed by dietary levels of 50 mg kg^{-1} of astaxanthin (Menasveta et al. 1993). Also, survival and growth rates of post-larvae increase according to dietary astaxanthin in *Penaeus monodon* (Merchie et al. 1998) and in *L. vannamei* up to supplementation levels of 200 and 400 mg kg^{-1} (Niu et al. 2009). Astaxanthin derived primarily from organic sources, such as organic crustacean shells, may be used in the feed ration for organic salmon and trout within the limit of their physiological needs. If organic sources are not available, natural sources of astaxanthin (such as Phaffia yeast) may be used, EC 889/2008.

Nutritional Strategies for a Make-Up of Feeds

Developing a better understanding of the molecular mechanisms controlling nutrient utilisation will help us to generate sustainable and functional diets and improve the efficiency of organic and conventional aquaculture (Mente et al. 2017). Proteomic studies have examined protein responses to dietary stimulations and provided information of the relationships between diet composition, protein metabolism and nutrient utilisation in aquatic animals (Rodrigues et al. 2012, 2018). Research has greatly enhanced knowledge of the metabolic pathways influenced by dietary changes in both liver and muscle in rainbow trout, Atlantic salmon, gilthead sea bream and white sea bream (Carter et al. 2012; Mente et al. 2017). Changing dietary sources that farmed fish are now fed and understanding their nutritional physiology and identifying key proteins that function in response to the different dietary nutrients will aid in formulating new aquafeeds and provide better fish growth performance and less environmental impact in both organic and conventional aquaculture.

Microorganisms', mostly bacteria, major roles lay in the nutrition of the animal host through various metabolic processes and the protection of the host against other pathogenic microorganisms. These bacterial communities tend to be rather diverse, and, thus, the identification of the exact role of each bacterium remains a challenging task. Diet is a major factor driving the composition and metabolism of the gut microbiota, while gut microbiota is actively involved in nutrient assimilation and immunity of the host organism. A nutrient is required in the diet if endogenous production from precursors is absent or insufficient to meet physiological needs. Fish cannot synthesise various compounds required for their metabolism and growth and derive these essential nutrients from their diet. Nevertheless, it is still unknown whether, and how, the supply of these nutrients is regulated. Research recently aims to examine which are these symbiotic microorganisms in the fish gut that can be controlled by the ingested feed and how the diet induces changes in the structure and function of the fish intestinal microbiota that accompany changes in their nutrition/welfare and health. The fish gut microbial communities, the gastrointestinal tract microbiota (GIT), can contribute nutrients and energy to the host via the fermentation of non-digestible dietary components and maintain a balance with the fish's metabolism and immune system. Changes in diet composition can be seen as a perturbation factor for the GIT microbial communities of the host organism causing shifts in these microbial communities. However, the extent to which microbial function varies with host demand and the underlying mechanisms are largely unknown. Ideally, any diet substitution should impose the minimal gut bacterial changes, or, at least, if any bacterial shifts occur, these should be functionally similar to those before the substitution, or, in a more eco-friendly/sustainable view, the new bacterial communities should be functionally similar to the ones of the wild fish populations.

The knowledge of fish and crustaceans' gastrointestinal tract microbiota (GIT) has progressed considerably over the past three decades (Ringø et al. 2016), with studies examining both natural as well as commercially reared populations (Meziti et al. 2010, 2012; Kormas et al. 2014; Mente et al. 2016) and the effect of different diets on GIT communities and the impact on fish growth and health. Studies exist on the effect of the diet on the fish intestinal microbiota, structure and morphology (Dimitroglou et al. 2009; Desai et al. 2012) and the effect of dietary components on fish gut microbiota (Ringø et al. 2016; Gatesoupe et al. 2014). High bacterial richness, determined by 454 pyrosequencing, has been found in the gut of wild and organic reared sea bream (*Sparus aurata*) compared to the conventional reared ones (Kormas et al. 2014) and in wild-caught vs. domesticated black tiger shrimps (*Penaeus monodon*) (Rungrassamee et al. 2014). In order to clarify this, along with the prevailing metabolic pathways related to fish nutrition, further research is required on the metabolic capacities of these GIT microorganisms. A novel approach called "engineering approach" to improve animal performance is to artificially select upon microbiomes, and thus engineering evolved microbiomes with specific effects on their host. This method selects upon microbial communities indirectly through the host and leverages host traits that evolved to influence microbiomes. A predictive framework can be developed for microbiome engineering that organises research around principles of microbiome selection, quantitative genetics and microbial community-ecology. The unravelling of intricate host-microbe symbioses and identification of core microbiome functions through metagenomics are essential to our ability to use the benefits of a healthy microbiome to our advantage in aquaculture, as well as gain deeper understanding of bacterial roles in vertebrate health (reviewed by Tarnecki et al. 2017). The future of organic aquaculture nutrition will benefit from a better understanding of the nutritional strategies and the fish's gut/microbe interactions and gut microorganism's diversity to allow the production of top-quality aquafeeds.

Potential Feed Produced in Aquaponics

EU organic regulation does not allow the certification of aquaponics products up to present since they use fertilisers to grow hydroponically the plants and do not use soil for the plants cultivation. It is clearly stated in the EU rules that recirculating technology and hydroponics are not permitted in organic production. If however you produce your plants in soil, and feed the fish with organic fish feed, and use those fertilisers listed in Annex I (Reg. 889/08), you can have your plants certified organic. A further harmonisation of terms and definitions used in the EU organic aquaculture and organic plant culture would enhance the strengthening of the organic system and food security.

Sustainable Fisheries

Towards ensuring the sustainability of fisheries, different systems have been introduced at different levels, like the United Nations Convention on the Law of the Sea, the FAO Code of Conduct for Responsible Fisheries, the United Nations Fish Stocks Agreement and different regional organisations (FAO 2011). The Common Fisheries Policy (CFP) regulation (EU Reg. 1380/2013) has established that the maximum sustainable yield should be reached for the target stocks and fishery should be conducted preserving also ecosystem functioning and integrity. The maximum sustainable yield (MSY) is the theoretical largest amount of fish that can be harvested from a stock over time without reduction in population size. This is the management tool that EU has committed to reach within 2020 for all commercially harvested fish stocks. There are also several independent organisations working on fish stock assessments and giving advice, e.g. FAO (UN Food and Agriculture Organization), that publish comprehensive statistics and information in order to provide politicians and other decision-makers with facts. Scientific, Technical and Economic Committee for Fisheries (STECF) is the official scientific body of the European Commission that revises annually the assessments performed at the level of advice/management bodies. Research on fish stock assessments and fishery management are currently carried out by means of “fishing effort regulation” or “total allowable catch (TAC)” both at national and international level. Examples of such organisations are ICES (the International Council for the Exploration of the Sea), GFCM (General Fisheries Commission for the Mediterranean) that is instrumental in coordinating efforts by governments to effectively manage fisheries at regional level, ICCAT (International Commission for the Conservation of Atlantic Tunas), Regional Fisheries Management Organizations (RFMOs), IMARPE (Peru – Institute of Fisheries Research) and IFOP (Chile – Institute of Fisheries Research). Advice from ICES is the basis for fisheries management in the EU, Iceland and Norway, while advices and recommendations from GFCM are the basis for the fisheries management in the Mediterranean (EU and non-EU). The European Commission establishes the rules of fisheries management and approves the management plans produced by member states, which then have to comply with them; otherwise they risk an infringement procedure. Private standards and certification schemes are developed to contribute to sustainability and responsible fisheries management (FAO 2011). The international fishmeal and fish oil organisation (IFFO) is represented by the major European, South American and a few other countries fishmeal and fish oil producers. IFFO has developed their IFFO-RS standard for responsible sourcing of raw materials (IFFO 2010), and an increasing number of production plants are certified in this system. With regard to the certification of the available ingredients for aquafeed and other animal feeds available, it is interesting to note that the global supply of fishmeal and fish oil significantly outperforms other feed ingredient supply when it comes to the volumes of certified product available (http://www.iffo.net/iffo-rs-statistics).The industry can quote a volume of certified product supply that currently exceeds 40% with a continued upward trend that is supported by the use of Fisheries Improvement Projects (FIPs) as a mechanism that brings advances in fisheries management and progression in marine ecosystem and socio-economic sustainability. In this way, the developing industry is actually enhancing the marine environment.

Marine Stewardship Council (MSC) is an independent, global, non-profit organisation with certification and eco-labelling programmes for fisheries and sustainable seafood (http://www.msc.org/). The MSC set science-based standards, and the certification process is performed by an accredited third party to ensure independence. Friend of the Sea is a non-profit non-governmental organisation (NGO), whose mission is the conservation of the marine habitat. Friend of the Sea's mission is to promote international certification project for products originated from both sustainable fisheries and aquaculture.

There are also other certification schemes for fisheries products; however, even if the quantity of marine ingredients obtained from sustainable sources increases, the global supply of fishmeal and fish oil is no longer able to meet the increasing demand from an expanding aquaculture industry. Nevertheless, the aquaculture sector has done relevant progress in research for alternative ingredients including plant products (Gatlin et al. 2007; Hardy 2010).

Food for Thought

Sustainable organic food production includes best environmental practices, a high level of biodiversity, the preservation of natural resources, the application of high animal welfare standards and a production method in line with the preference of certain consumers for products produced using natural substances and processes. Feed for organic fish is currently produced according to the EU regulations 834/2007, 889/2008 and 710/2009, 1358/2014 and 673/2016. However, due to the limited amount of adequate options of ingredients to better match amino acid and fatty acid profiles and covering the dietary needs of other essential nutrients for the full organic production cycle, i.e. broodstock, fry, juveniles and for on-growing, it is challenging for the feed industry to comply with organic regulation (Table 8.3).

It is well defined that for organic fish and shrimp production, we need to supply dietary nutrients either through the exogenous supply of organic feed inputs (ranging from the use of organically produced and certified live and/or processed animal or plant feed items to the use of commercially formulated feeds composed of mix-

Table 8.3 Summary of the main current concerns and challenges arising from organic aquafeeds and feeding

Weakness	Approaches/planning steps to achieve sustainability	Challenges
• Increased pressure on feed-grade fisheries. Affects wild fish populations. Wider ecosystem impacts • Fish meal and fish oil replacement • Fish meal trimming from organic aquaculture trimmings or from fish already caught for human consumption	• Substitution by alternative organic ingredients to develop nutritionally efficient diets • Reduce feed conversion ratio • Microbial floc meal/microalgae species • Use probiotics in feed • Enhancement of beneficial microfauna and flora • Better adapted stock • Increase organic aquaculture sources	• Determine the nutrient bioavailability of organic aquafeeds • Better understanding of the role of effective microorganisms. Gut microflora affects digestibility and nutrient utilisation and excretion • Culturing microalgae by fermentation • Improve manufacturing process of Fish meal trimmings
• Regulate feed intake • Lower stocking densities, lower production volume, longer production cycles • Improve diet formulations to enhance growth and reproductive performance	• Feeding strategies. Control naturally physiological processes that regulate feeding behaviour and appetite (hypothalamic factors) • Improve feed palatability, quality and nutrient availability, inactive anti-nutritional factors • Determine the physiological basis, which enhances expression of growth of species • Evaluate nutritional effects on growth and reproductive efficiency, gonadal and larval quality • Polyculture, sea ranching • Integrated multi-trophic Aquaculture (IMTA)	• Optimise the utilisation of feed resources • Evaluate the performance of new organic diet formulations with regard to feed consumption and growth • Major focus of aquatic animal nutrition research to improve diet formulation to enhance growth performance in aquaculture
• Establishment of optimal dietary requirements of amino acids and supplement diets with no synthetic amino acids	• Optimise the dietary amino acid profile of the feed • Characterisation of alternative protein/amino acid sources • Essential amino acids produced through fermentation • Microbial supply of amino acids scaled to its demand	• Feed should cover the aquatic animal's needs for amino acids according to the organic principles • Use models to understand tissue growth and nutrition on flesh quality and composition • Amino acids produced by fermentation allows replacement of high-quality Fish meal • Aquatic animal's gut microbiota can in theory play a critical role in providing the necessary nutrients

• Understanding of functional genomics associated with growth, development and nutrition • Optimise reproductive performance, growth, nutrient retention and product quality	• Examine changes in gene expression in response to alterations in feed nutrients • Better understanding of the physiological and behavioural needs of the animals	• Identify mechanisms by which nutrients affect or regulate genes for muscle growth • Increase knowledge of nutritional strategies and cellular functions affecting reproductive physiology
• Update regularly legislation due to recent research findings on nutrition	• Monitoring and enforcement • Actively incorporate the research nutrition results	• Research and policy development • Fund more nutritional research experiments • Enhance the research progress

Adopted and modified from Mente et al. (2011)

tures of different organically certified feed ingredient sources blended so as to satisfy the precise dietary nutrient requirements of the cultured species) or through the endogenous production of live food organisms (including plankton, benthic organisms, biofloc, etc.) within the culture system for the target species. Simple as this statement is, most commercially formulated feeds currently used to produce organically certified finfish (including salmonids, sea bass, seabream, etc.) and crustaceans (shrimp) rely on a significant proportion of feed ingredients derived from wild sustainable capture fisheries (fishmeal, fish oil). These wild fisheries are stagnated. There has been considerable pressure by the commercial aquaculture feed manufacturing sector and organic aquaculture certification bodies to assure an appropriate level of the above feed ingredient sources, as they are natural sources of feeds for these higher-value carnivorous fish and crustacean species. However, as within humans and terrestrial farm animals, fish and shrimp do not have a dietary requirement for fishmeal or fish oil but rather have a dietary requirement for the essential nutrients contained within these feed ingredient sources (Tacon and Metian 2015). Furthermore, in a more eco-friendly/sustainable view, the better understanding of how intestinal microbiota can contribute nutrients and energy to the fish for better growth and health is one of the key research topics for organic aquaculture development.

The ongoing positive growth trend of the aquaculture industry is expected to continue, reflecting the rising demand for healthy human food products. Hence, since 2000 there has been an increasing demand for seafood that has been farmed according to certified organic standards, notably in European countries. Organic certification relates to the production of organic food products such as fish and shrimp in a protected environment where we can control feed inputs, the quality of the environment and the husbandry practices. It follows, therefore, that like the terrestrial organic livestock production sector, the onus is on the aquatic nutritionist to formulate an organic feed close to the feed that each fish and shrimp species are consuming in their natural environment. The organic feed formulation could use certified organic feed ingredient sources to stimulate bacterial communities to be functionally similar to the ones of the wild fish populations in order to meet the essential dietary nutrient requirements of the target species.

References

Aas TS, Hatlen B, Grisdale-Helland B, Terjesen BF, Bakke-McKellep AM, Helland SJ (2006) Effects of diets containing a bacterial protein meal on growth and feed utilisation in rainbow trout (*Oncorhyncus mykiss*). Aquaculture 261:357–368

Adesulu EA, Mustapha AK (2000) Use of housefly maggots as a fishmeal replacer in tilapia culture: a recent vogue in Nigeria. In: Fitzsimmons K, Filho JC (eds) 5th International Symposium on Tilapia Aquaculture, Rio de Janeiro, Brazil, , vol 1. p 138

Ajani EK, Nwanna LC, Musa BO (2004) Replacement of fishmeal with maggot meal in the diets of Nile tilapia, *Oreochromis niloticus*. World Aquac 35(1):52–54

Albrektsen S, Hope B, Aksnes A (2009) Phosphorous (P) deficiency due to low P availability in fishmeal produced from blue whiting (*Micromesistius poutassou*) in feed for under-yearling Atlantic salmon (*Salmo salar*) smolt. Aquaculture 296:318–328

Alday-Sanz V (2011) The shrimp book. Nottingham University Press, Nottingham, p 930. ISBN-10: 1904761593

Alexis MN, Karanikolas KK, Richards RH (1997) Pathological findings owing to the lack of ascorbic acid in cultured gilthead bream (*Sparus aurata* L.). Aquaculture 151:209–218

Altan Ö, Korkut AY (2011) Appearent digestibility of plant protein based diets by European Sea bass *Dicentrarchus labrax* L., 1758. Turk J Fish Aquat Sci 11:87–92

Ambasankar K, Ahmad SA, Dayal JS (2006) Effect of dietary phosphorus on growth and its excretion in Tiger shrimp, *Penaeus monodon*. Asian Fish Sci 19:21–26

Aniebo AO, Owen OJ (2010) Effects of age and method of drying on the proximate composition of Housefly Larvae (*Musca domestica* Linnaeus) meal (HFLM). Pak J Nutr 9:485–487

Anupama, Ravindra P (2000) Value-added food: single cell protein. Biotechnol Adv 6:459–479

Atalah E, Cruz CMH, Izquierdo MS, Rosenlund G, Caballero MJ, Valencia A, Robaina L (2007) Two microalgae *Crypthecodinium cohnii* and *Phaeodactylum tricornutum* as alternative source of essential fatty acids in starter feeds for seabream (*Sparus aurata*). Aquaculture 270:178–185

Azevedo PA, Cho CY, Bureau DP (1998) Effects of feeding level and water temperature on growth, nutrient and energy utilization and waste outputs of rainbow trout (*Oncorhynchus mykiss*). Aquat Living Resour 11:227–238

Baron C, Svendsen G, Lund I, Jokumsen A, Nielsen H, Jacobsen C (2013) Organic plant ingredients in the diet of Rainbow Trout (*Oncorhynchus mykiss*). Impact on fish muscle composition and oxidative stability. Eur J Lipid Sci Technol 115:0000–0000. https://doi.org/10.1002/ejlt.201300157.

Bell MV, Dick JR (2004) Changes in capacity to synthesise 22:6n-3 during early development in rainbow trout (*Oncorhynchus mykiss*). Aquaculture 235:393–409

Bell JG, Koppe W (2011) Lipids in aquafeeds. In: Turchini GM, Ng W-K, Tocher DR (eds) Fish oil replacement and alternative lipid sources in aquaculture feeds. Taylor & Francis, CRC Press, Boca Raton, pp 21–59

Bell JG, Tocher DR, Farndale BM, Cox DI, McKinney RW, Sargent JR (1997) The effect of dietary lipid on polyunsaturated fatty acid metabolism in Atlantic salmon (*Salmo salar*) undergoing Parr-Smolt transformation. Lipids 32(5):515–525

Bell MV, Dick JR, Porter AEA (2001) Biosynthesis and tissue deposition of docosahexaenoic acid (22:6n-3) in rainbow trout (*Oncorhynchus mykiss*). Lipids 36:1153–1159

Bendiksen EA, Johnsen CA, Olsen HJ, Jobling M (2011) Sustainable aquafeeds: progress towards reduced reliance upon marine ingredients in diets for farmed Atlantic salmon (*Salmo salar L.*). Aquaculture 314:132–139

Berge GM, Baeverfjord G, Skrede A, Storebakken T (2005) Bacterial protein grown on natural gas as protein source in diets for Atlantic salmon, *Salmo salar*, in saltwater. Aquaculture 244:233–240

Bergleiter S, Berner U, Censkowsky U, Julia-Camprodon G (2009) Organic aquaculture 2009. Production and markets. Naturland e.V. and Organic Services GmbH, Munich. Report, 120 pages

Bondari K, Sheppard DC (1987) Soldier fly *Hermetia illucens* L., as feed for channel catfish, Ictalurus punctatus (Rafinesque) and blue tilapia, (*Oreochromis aureus*) (Steindachner). Aquac Fish Manag 18:209–220

Borquez A, Serrano E, Dantagnan P, Carrasco J, Hernandez A (2011) Feeding high inclusion of whole grain white lupin (*Lupinus albus*) to rainbow trout (*Oncorhynchus mykiss*): effects on growth, nutrient digestibility, liver and intestine histology and muscle fatty acid composition. Aquac Res 42:1067–1078

Breck O, Bjerkas E, Campbell P, Arnesen P, Haldorsen P, Waagbø R (2003) Cataract preventative role of mammalian blood meal, histidine, iron and zinc in diets for Atlantic salmon (*Salmo salar*) of different strains. Aquac Nutr 9:341–350

Brown MR, Mular M, Miller I, Farmer C, Trenerry C (1999) The vitamin content of microalgae used in aquaculture. J Appl Phycol 11:247–255

Bureau DP, Harris AM, Cho CY (1999) Apparent digestibility of rendered animal protein ingredients for rainbow trout (*Oncorhynchus mykiss*). Aquaculture 180:345–358

Buzzi M, Henderson RJ, Sargent JR (1996) The desaturation and elongation of acid and eicosapentaenoic acid by hepatocytes and liver microsomes from rainbow trout (*Oncorhynchus mykiss*) fed diets containing fish oil or olive oil. Biochim Biophys Acta 1299:35–244

Cahu CL, Infante JLZ, Peres A, Quazuguel P, Le Gall MM (1998) Algal addition in sea bass (*Dicentrarchus labrax*) larvae rearing: effect on digestive enzymes. Aquaculture 161:479–489

Carter CG, Hauler RC (2000) Fish meal replacement by plant meals in extruded feeds for Atlantic salmon, *Salmo salar* L. Aquaculture 185:299–311

Carter CG, Houlihan DF (2001) Protein synthesis. In: Wright PA, Anderson PM (eds) Nitrogen excretion, fish physiology, vol 19. Academic, New York, pp 31–75

Carter CG, Sajjadi M (2011) Low fishmeal diets for Atlantic salmon, *Salmo salar* L. using soy protein concentrate treated with graded levels of phytase. Aquac Int 19:431–444

Carter CG, Katersky RS, Barnes J, Hauler RC, Bridle AR (2010) Redefining nutrient requirements of fish in sub-optimum environments. In: Crovetto GM (ed) Energy and protein metabolism and nutrition, EAAP Publication 127. Wageningen Academic Publishers, Wageningen, pp 445–446

Carter CG, Mente E, Katersky Barnes RS, Nengas I (2012) Protein synthesis in gilthead sea bream: response to partial fishmeal replacement. Br J Nutr 108:2190–2197

Chebbaki K, Akharbach H, Talbaoui E, Abrehouch A, Ait A, Sedki S, Ben Bani A, Idaomar M (2010) Effect of fish meal replacement by protein sources on the extruded and pressed diet of European sea bass juvenile (*Dicentrarchus labrax*). Agric Biol J N Am 1:704–710

Chen HY (1993) Requirements of marine shrimp, *Penaeus monodon*, juveniles for phosphatidylcholine and cholesterol. Aquaculture 109:165–176

Chen HY (1998) Nutritional requirements of the black tiger shrimp: *Penaeus monodon*. Rev Fish Sci 6(1):79–95

Choubert G, Mendes-Pinto MM, Morais R (2006) Pigmenting efficacy of astaxanthin fed to rainbow trout *Oncorhynchus mykiss*: effect of dietary astaxanthin and lipid sources. Aquaculture 257:429–436

Codabaccus BM, Bridle AR, Nichols PD, Carter CG (2011) Effect of feeding Atlantic salmon (*Salmo salar* L.) a diet enriched with stearidonic acid from parr to smolt on growth and *n-3* LC-PUFA biosynthesis. Br J Nutr 105:1772–1782

Conklin DE (2003) Use of soybean meal in the diets of marine shrimp, Technical review paper AQ 144–2003. Department of Animal Science, University of California, Davis, United Soybean Board and American Soybean Association, Davis

Cottee SY, Petersan P (2009) Animal welfare and organic aquaculture in open systems. J Agric Environ Ethics 22:437–461

Cousin M (1995) Contribution à l'étude de utilisation des glucides et du rapport proteine/énergie chez *P. vannamei* et *P. stylrostris*. Thèse INA/PG. Paris. p 201

Coutinho P, Rema P, Otero A, Pereira O, Fabregas J (2006) Use of biomass of the marine microalga *Isochrysis galbana* in the nutrition of goldfish (*Carassius auratus*) larvae as source of protein and vitamins. Aquac Res 37:793–798

Crampton VO, Nanton DA, Ruohonen K, Skjervold PO, El-Mowafi A (2010) Demonstration of salmon farming as a net producer of fish protein and oil. Aquac Nutr 16:437–446. https://doi.org/10.1111/j.1365-2095.2010.00780.x

Cruz-Suárez LE, Antimo-Pérez JS, Luna-Mendoza N, Tapia-Salazar M, Guajardo-Barbosa C, Ricque-Marie D (2000) Relaciones proteína/energía y proteína vegetal/animal optimas en alimentos de engorda para *Litopenaeus vannamei* y *L. stylirostris*. In: Cruz-Suárez LE, Ricque-Marie D, Tapia-Salazar M, Olvera-Novoa MA, Civera-Cerecedo R (eds) Avances en Nutrición Acuícola V. Memorias del V Simposium Internacional de Nutrición Acuícola. 19–22 Noviembre, 2000. Mérida

Cummins VC Jr, Rawles SD, Thompson KR, Velasquez A, Kobayashi Y, Hager J, Webster CD (2017) Evaluation of black soldier fly (*Hermetia illucens*) larvae meal as partial or total replacement of marine fish meal in practical diets for pacific white shrimp (*Litopenaeus vannamei*). Aquaculture 473:337–344. https://doi.org/10.1016/j.aquaculture.2017.02.022

Dallaire V, Lessard P, Vandenberg G, de la Noue J (2007) Effect of algal incorporation on growth, survival and carcass composition of rainbow trout (*Oncorhynchus mykiss*) fry. Bioresour Technol 98:1433–1439

Davis DA, Lawrence AL, Gatlin DM (1993) Response of *Penaeus vannamei* to dietary calcium, phosphorus and calcium: phosphorus ratio. J World Aquacult Soc 24(4):504–515

Dayal JS, Ali SA, Ambasankar K, Singh P (2003) Effect of dietary protein level on its in vitro and in vivo digestibility in the tiger shrimp *Penaeus monodon* (Crustacea): Penaeidae. Indian J Mar Sci 32(2):151–155

De Francesco M, Parisi G, Pérez-Sanchez J, Gomez-Réqueni P, Médale F, Kaushik SJ, Mecatti M, Poli BM (2007) Effect of high-level fish meal replacement by plant proteins in gilthead sea bream (*Sparus aurata*) on growth and body/fillet quality traits. Aquac Nutr 13:361–372

Desai AR, Links MG, Collins SA, Mansfield GS, Drew MD, Van Kessel AG et al (2012) Effects of plant-based diets on the distal gut microbiome of rainbow trout (*Oncorhynchus mykiss*). Aquaculture 350–353:134–142

Devic E, Leschen W, Murray F, Little DC (2018) Growth performance, feed utilization and body composition of advanced nursing Nile tilapia (*Oreochromis niloticus*) fed diets containing black soldier fly (*Hermetia illucens*) larvae meal. Aquac Nutr 24:416–423

Dimitroglou A, Merrifield DL, Moate R, Davies SJ, Spring P, Sweetman J et al (2009) Dietary mannanoligosaccharide supplementation modulates intestinal microbial ecology and improves gut morphology of rainbow trout, *Oncorhynchus mykiss* (Walbaum). J Anim Sci 87:3226–3234

Domínguez D, Rimoldi S, Robaina LE, Torrecillas S, Terova G, Zamorano MJ, Karalazos V, Hamre K, Izquierdo M (2017) Inorganic, organic, and encapsulated minerals in vegetable meal based diets for *Sparus aurata* (Linnaeus, 1758). PeerJ 5:e3710. https://doi.org/10.7717/peerj.3710

EC 834/2007 Council Regulation (EC) No 834/2007 of 28 June 2007 on organic production and labelling of organic products and repealing Regulation (EEC) No 2092/91

EC 889/2008 Commission Regulation (EC) No 889/2008 of 5 September 2008 laying down detailed rules for the implementation of Council Regulation (EC) No 834/2007 on organic production and labelling of organic products regarding organic production, labelling and control

EC 710/2009 Commission Regulation amending Regulation (EC) No 889/2008 laying down detailed rules for the implementation of Council Regulation (EC) No 834/2007, as regards laying down detailed rules on organic aquaculture animal and seaweed production. Official Journal of the European Union

EC 1380/2013 Regulation (EU) No 1380/2013 of the European Parliament and of the Council of 11 December 2013 on the Common Fisheries Policy, amending Council Regulations (EC) No 1954/2003 and (EC) No 1224/2009 and repealing Council Regulations (EC) No 2371/2002 and (EC) No 639/2004 and Council Decision 2004/585/EC

EC 1358/2014 Comm. Reg.- No 1358/2014 of 18 December 2014 amending Regulation (EC) No 889/2008 laying down detailed rules for the implementation of Council Regulation (EC) No 834/2007 as regards the origin of organic aquaculture animals, aquaculture husbandry practices, feed for organic aquaculture animals and products and substances allowed for use in organic aquaculture

EC 673/2016 Commission Implementing Regulation (EU) 2016/673 of 29 April 2016 amending Regulation (EC) No 889/2008 laying down detailed rules for the implementation of Council Regulation (EC) No 834/2007 on organic production and labelling of organic products with regard to organic production, labelling and control

EFSA – European Food Safety Authority (2011) Scientific opinion on the revision of the quantitative risk assessment (QRA) of the BSE risk posed by processed animal proteins (PAPs). EFSA J 9(1):1947

El-Haroun ER, Bureau DP (2007) Comparison of bioavailability of lysine in blood meals of various origins to that of L-lysine HCl for rainbow trout (*Oncorhynchus mykiss*). Aquaculture 262:402–409

El-Nawwi SA, El-Kader AA (1996) Production of single-cell protein and cellulase from sugarcane bagasse: effect of culture factors. Biomass Bioenergy 11:361–364

Emerenciano M, Ballester EL, Cavalli RO, Wasielesky W (2012) Biofloc technology application as a food source in a limited water exchange nursery system for pink shrimp *Farfantepenaeus brasiliensis* (Latreille, 1817). Aquac Res 43(3):447–457

Enami HR (2011) A review of using canola/rapeseed meal in aquaculture feeding. J Fish Aquat Sci 6:22–36

Eroldogan T, Turchini GM, Yilmaz AH, Tasbozan O, Engin K, Ölçülü A, Özsahinoglu I, Mumogullarinda P (2012) Potential of cottonseed oil as fish oil replacer in European Sea Bass feed formulation. Turk J Fish Aquat Sci 12:787–797

Espe M, Lemme A, Petri A, El-Mowafi A (2006) Can Atlantic salmon (*Salmo salar*) grow on diets devoid of fish meal? Aquaculture 255:255–262

FAO (2011) Private standards and certification in fisheries and aquaculture. Current practice and emerging issues, FAO Fisheries and Aquaculture Technical paper. No. 553. Food and Agriculture Organisation of the United Nations (FAO), Rome, p 181

Fasakin EA, Balogun AM, Ajayi OO (2003) Evaluation of full fat and defatted maggot meals in the feeding of clariid catfish *Clarias gariepinus* fingerlings. Aquac Res 34:733–738

Figueiredo-Silva A, Rocha E, Dias J, Silva P, Rema P, Gomes E, Valente LMP (2004) Partial replacement of fish oil by soybean oil on lipid distribution and liver histology in European sea bass (*Dicentrarchus labrax*) and rainbow trout (*Oncorhynchus mykiss*) juveniles. Aquac Nutr 11:147–155

Fountoulaki EE, Alexis MN, Nengas I (2005) Protein and energy requirements of gilthead bream (*Sparus aurata*, L.) fingerlings: preliminary results. In: Montero D, Basurco B, Nengas I, Alexis M, Izquierdo M (eds) Mediterranean fish nutrition, CIHEAM Cahiers Options Méditerranéennes 63. CIHEAM, Zaragoza, pp 19–26

Fournier V, Gouillou-Coustans MF, Kaushik SJ (2000) Hepatic ascorbic acid saturation is the most stringent response criterion for determining the vitamin C requirement of juvenile European sea bass (*Dicentrarchus labrax*). J Nutr 130:617–620

Fox J, Treece GD, Sanchez D (1998) Shrimp nutrition and feed management. Methods for improving shrimp farming in Central America. Univ Centroam Cent Investig Ecosistemas Acuáticos 2001:65–90

Francis G, Makkar HPS, Becker K (2001) Antinutritional factors present in plant derived alternative fish feed ingredients and their effects in fish. Aquaculture 199:197–227

Freuchtnicht GW, Bark LE, Malecha SR, Stanley RW (1988) The effect of protein level in the feed on growth performance of fresh water prawn *Macrobrachium rosenbergii* individually recorded in clear water flow through aquaria. Presented at the 19th annual meeting of the world aquaculture society, Honolulu, Hawaii

Fujii K, Imazato E, Nakashima H, Ooi O, Saeki A (2006) Isolation of the non-fastidious microalga with astaxanthin-accumulating property and its potential for application to aquaculture. Aquaculture 261:285–293

Ganuza E, Benitez-Santana T, Atalah E, Vega-Orellana O, Ganga R, Izquierdo MS (2008) *Crypthecodinium cohnii* and *Schizochytrium sp* as potential substitutes to fisheries-derived oils from seabream (*Sparus aurata*) microdiets. Aquaculture 277:109–116

Gasco L, Henry M, Piccolo G, Marono S, Gai F, Renna M, Lussiana C, Antonopoulou E, Mola P, Chatzifotis S (2016) *Tenebrio molitor* meal in diets for European sea bass (*Dicentrarchus labrax* L.) juveniles: growth performance, whole body composition and in vivo apparent digestibility. Anim Feed Sci Technol 220:34–45

Gasco L, Gai F, Maricchiolo G, Genovese L, Ragonese S, Bottari T, Caruso G (2018) Fishmeal alternative protein sources for aquaculture feeds. In: Gasco L, Gai F, Maricchiolo G, Genovese

L, Ragonese S, Bottari T, Caruso G (eds) Feeds for the aquaculture sector: current situation and alternative sources. Springer, Cham, pp 1–28

Gatesoupe FJ, Huelvan C, Le Bayon N, Severe A, Aasen IM, Degnes KF et al (2014) The effects of dietary carbohydrate sources and forms on metabolic response and intestinal microbiota in sea bass juveniles, *Dicentrarchus labrax*. Aquaculture 422:47–53

Gatlin DM, Barrows FT, Brown P, Dabrowski K, Gaylord TG, Hardy RW, Herman E, Hu GS, Krogdahl A, Nelson R, Rust M, Sealey W, Skonberg D, Souza EJ, Stone D, Wilson R, Wurtele E (2007) Expanding the utilization of sustainable plant products in aquafeeds; a review. Aquac Res 38:551–579

Gatta PP, Pirini M, Testi S, Vignola G, Monetti PG (2000) The influence of different levels of dietary vitamin E on sea bass *Dicentrarchus labrax* flesh quality. Aquac Nutr 6:47–52

Geay F, Santigosa I, Culi E, Corporeau C, Boudry P, Dreano Y, Corcos L, Bodin N, Vandeputte M, Zambonino-Infante JL, Mazurais D et al (2010) Regulation of FADS2 expression and activity in European sea bass (*Dicentrarchus labrax*, L.) fed a vegetable diet. Comp Biochem Physiol B 156:237–243

Geay F, Ferraresso S, Zambonino Infante JL, Bargelloni L, Quentel C, Vandeputte M, Kaushik SJ, Cahu C, Mazurais C (2011) Effects of the total replacement of fish-based diet with plant-based diet on the hepatic transcriptome of two European sea bass (*Dicentrarchus labrax*) half-sibfamilies showing different growth rates with the plant-based diet. BMC Genomics 12:522–539

Gensch C-O, Graulich K, Teufel J (2004) Ecostudy of Lucanthin® Pink for Salmon production. BASF, Ludwigshafen. 23pp

Glencross BD, Smith DM, Thomas MR, Williams KC (2002) The effects of dietary lipid amount and fatty-acid composition on the digestibility of lipids by the prawn, *Penaeus monodon*. Aquaculture 205(1–2):157–169

Glencross B, Rutherford N, Hawkins W (2011) A comparison of the growth performance of rainbow trout (*Oncorhynchus mykiss*) when fed soybean, narrow-leaf or yellow lupin meals in extruded diets. Aquac Nutr 17:317–325

Gomes da Silva J, Oliva Teles A (1998) Apparent digestibility of feedstuffs in seabass (*Dicentrarchus labrax*) juveniles. Aquat Living Resour 11:187–191

Gong H, Lawrence AL, Jiang D-H, Castille FL, Gatlin DMI (2000) Lipid nutrition of juvenile *Litopenaeus vannamei* I. Dietary cholesterol and de-oiled soy lecithin requirements and their interaction. Aquaculture 190:305–324

Grisdale-Helland B, Ruyter B, Rosenlund G, Obach A, Helland SJ, Sandberg MG, Standal H, Røsjø C (2002) Influence of high contents of dietary soybean oil on growth, feed utilization, tissue fatty acid composition, heart histology and standard oxygen consumption of Atlantic salmon (*Salmo salar*) raised at two temperatures. Aquaculture 207:311–329

Grisdale-Helland B, Gatlin DM, Corrent E, Helland SJ (2011) The minimum dietary lysine requirement, maintenance requirement and efficiency of lysine utilization for growth of Atlantic salmon smolts. Aquac Res 42:1509–1529

Grisdale-Helland B, Lemme A, Helland SJ (2013) Threonine requirement for maintenance and efficiency of utilization for threonine accretion in Atlantic salmon smolts determined using increasing ration levels. Aquaculture 372–375:158–166

Hale OM (1973) Dried *Hermetia illucens* larvae (Stratiomyidae) as a feed additive for poultry. J Georgia Entomol Soc 8:16–20

Hardy RW (2010) Utilization of plant proteins in fish diets: effects of global demand and supplies of fishmeal. Review article. Aquacult Res 41:770–776

Hardy RW, Barrows FT (2002) Diet formulation and manufacture. In: Halver JE, Hardy RW (eds) Fish nutrition, 3rd edn. Academic, London, pp 505–600

He H, Lawrence AL (1993a) Vitamin C requirements of the shrimp *Penaeus vannamei*. Aquaculture 114(3):305–316

He H, Lawrence AL (1993b) Vitamin E requirement of *Penaeus vannamei*. Aquaculture 118(3):245–255

Henrique MMF, Gomes EF, Gouillou-Coustans MF, Oliva-Teles A, Davies SJ (1998) Influence of supplementation of practical diets with vitamin C on growth and response to hypoxic stress of seabream, *Sparus aurata*. Aquaculture 161:415–426

Henry M, Gasco L, Piccolo G, Fountoulaki E (2015) Review on the use of insects in the diet of farmed fish: past and future. Anim Feed Sci Technol 203:1–22

Hidalgo F, Alliot E (1988) Influence of water temperature on protein requirement and protein utilization in juvenile sea bass, *Dicentrarchus labrax*. Aquaculture 72:115–129

Hixon SM, Parrish CC, Anderson DM (2014) Full substitution of fish oil with camelina (*Camelina sativa*) oil, with partial substitution of fish meal with camelina meal, in diets for farmed Atlantic salmon (*Salmo salar*) and its effects on tissue lipids and sensory quality. Food Chem 157:51–61

Iaconisi V, Marono S, Parisi G, Gasco L, Genovese L, Maricchiolo G, Bovera F, Piccolo G (2017) Dietary inclusion of *Tenebrio molitor* larvae meal: effects on growth performance and final quality treats of blackspot sea bream (*Pagellus bogaraveo*). Aquaculture 476:49–58

IFFO (2010) Global standard for responsible supply. Requirements for certification. International Fishmeal and Fish oil Organisation (IFFO), p 24. 1st November 2010

Izquierdo MS, Obach A, Arantzamendi L, Montero D, Robaina L, Rosenlund G (2003) Dietary lipid sources for seabream and seabass: growth performance, tissue composition and flesh quality. Aquacult Nutr 9:397–407

Kanazawa A, Teshima S, Sakamoto M (1985) Effects of dietary lipids, fatty acids and phospholipids on growth and survival of prawn (Peaneus Japonicus) larvae. Aquaculture 50:39–49

Karapanagiotidis IT, Daskalopoulou E, Vogiatzis I, Rumbos C, Mente E, Athanassiou CG (2014) Substitution of fishmeal by fly *Hermetia illucens* prepupae meal in the diet of gilthead seabream (*Sparus aurata*). HydroMedit 2014:110–114

Karapanagiotidis IT, Psofakis P, Mente E, Malandrakis E, Golomazou E (2018) Effect of fishmeal replacement by poultry by-product meal on growth performance, proximate composition, digestive enzyme activity, haematological parameters and gene expression of gilthead seabream (*Sparus aurata*). Aquacult Nutr, 1–12

Kaushik SJ, Seiliez I (2010) Protein and amino acid nutrition and metabolism in fish: current knowledge and future needs. Aquac Res 41:322–332

Kaushik SJ, Gouillou-Coustans MF, Cho CY (1998) Application of the recommendations on vitamin requirements of finfish by NRC (1993) to salmonids and sea bass using practical and purified diets. Aquaculture 161:463–474

Kaushik SJ, Covès D, Dutto G, Blanc D (2004) Almost total replacement of fish meal by plant protein sources in the diet of a marine teleost, the European seabass, *Dicentrarchus labrax*. Aquaculture 230:391–404

Kissil GW, Cowey CB, Adron JW, Richards RH (1981) Pyridoxine requirements of the gilthead bream, *Sparus aurata*. Aquaculture 23:243–255

Kjær MA, Todorcevic M, Torstensen BE, Vegusdal A, Ruyter B (2008) Dietary n-3 HUFA affects mitochondrial fatty acid β-oxidation capacity and susceptibility to oxidative stress in Atlantic Salmon. Lipids 43(9):813–827

Kontara EKM, Coutteau P, Sorgeloos P (1997) Effect of dietary phospholipid on requirements for and incorporation of n − 3 highly unsaturated fatty acids in postlarval *Penaeus japonicus* Bate. Aquaculture 158:305–320

Kormas KA, Meziti A, Mente E, Fretzos A (2014) Gut microorganisms in organic and conventional sea bream. Microbiol Open 3:718–728. https://doi.org/10.1002/mbo3.202

KRAV (2009) Standards for KRAV-certified production – January 2009. www.krav.se

KRAV (2010) Standards for Krav-certified production, 2010. kap 5: Djurhållning. http://www.krav.se/KravRegler

Krogdahl A, Penn M, Thorsen J, Refstie S, Bakke AM (2010) Important anti-nutrients in plant feedstuffs for aquaculture: an update on recent findings regarding responses in salmonids. Aquac Res 41:333–344

Kuhn DD, Boardman GD, Lawrence AL, Marsh L, Flick GJ (2009) Microbial floc meals as a replacement ingredient for fish meal and soybean protein in shrimp feed. Aquaculture 296:51–57

Kureshy N, Davis DA (2000) Metabolic requirement for protein by pacific white shrimp, *Litopenaeus vannamei*. Memorias del V symposium Internacional de Nutricion Acuicola. 19–22 Noviembre, 2000. Mérida

Lanari D, D'Agaro E (2005) Alternative plant protein sources in sea bass diets. Ital J Anim Sci 4:365–374

Lands WEM (1992) Biochemistry and physiology of n-3 fatty acids. FASEB J 6(8):2530–2536

Larsen BK, Dalsgaard J, Pedersen PB (2012) Effects of plant proteins on postprandial, free plasma amino acid concentrations in rainbow trout (*Oncorhynchus mykiss*). Aquaculture 326–329:90–98

Lazzari R, Baldisserotto B (2008) Nitrogen and phosphorus waste in fish farming. B Inst Pesca São Paulo 34(4):591–600

Lee MH, Shiau SY (2002) Dietary vitamin C and its derivates affect immune response in grass shrimp, *Penaeus monodon*. Fish Shellfish Immunol 12:119–129

Lee MH, Shiau SY (2004) Vitamin E requirements of juvenile grass shrimp, *Penaeus monodon*, and effects on non-specific immune responses. Fish Shellfish Immunol 16(4):475–485

Li P, Gatlin DM (2003) Evaluation of brewers yeast (*Saccharomyces cerevisiae*) as a feed supplement for hybrid striped bass (*Morone chrysops x M. saxatilis*). Aquaculture 219:681–692

Lund I, Dalsgaard J, Rasmussen HT, Holm J, Jokumsen A (2011) Replacement of fish meal with a matrix of organic plant proteins in organic trout (*Oncorhynchus mykiss*) feed, and the effects on nutrient utilization and fish performance. Aquaculture 321:259–266. https://doi.org/10.1016/j.aquaculture.2011.09.028

Lund I, Dalsgaard J, Hansen JH, Jacobsen C, Holm J, Jokumsen A (2012) Effects of organic plant oils and role of oxidation on nutrient utilization in juvenile rainbow trout (*Oncorhynchus mykiss*). Int J Anim Biosci 1–10. doi:https://doi.org/10.1017/S1751731112001693

Magalhães R, Sanchez-Lopez A, Leal RS, Martınez-Lorens S, Oliva-Teles A, Peres H (2017) Black soldier fly (*Hermetia illucens*) prepupae meal as a fish meal replacement in diets for European seabass (*Dicentrarchus labrax*). Aquaculture 476:79–85. https://doi.org/10.1016/j.aquaculture.2017.04.021

Mahmood SH, Shahadat AM, Hossain ML (2005) Growth of black tiger shrimp, *Penaeus monodon*, on fishmeal based formulated diet in a Southeastern Coastal shrimp farm of Bangladesh. Pak J Zool 37:95–100

Mai KS, Zhang L, Ai QH, Duan QY, Li HT, Wan JL, Liufu ZG (2006) Dietary lysine requirement of juvenile seabass (*Lateolabrax japonicas*). Aquaculture 258:535–542

Mathews JA, Tan H, Moore MJB, Bell G (2011) A conceptual lignocellulosic 'feed+fuel' biorefinery and its application to the linked biofuel and cattle raising industries in Brazil. Energy Policy 39:4932–4938

McVey JP (1993) CRC handbook of mariculture, Crustacean aquaculture, vol I, 2nd edn. CRC Press, Boca Raton. 526 pp

Menasveta P, Worawattanamateekul W, Latscha T, Clark JS (1993) Correction of black tiger prawn (*Penaeus monodon fabricius*) coloration by astaxanthin. Aquac Eng 12(4):203–213

Mente E, Karalazos V, Karapanagiotidis IT, Pita C (2011) Nutrition in organic aquaculture: an inquiry and a discourse. Aquac Nutr 17:798–817

Mente E, Stratakos A, Boziaris IS, Kormas KA, Karalazos V, Karapanagiotidis IT, Catsiki AV, Leondiadis L (2012) The effect of organic and conventional production methods on sea bream growth, health and body composition; a field experiment. Sci Mar. https://doi.org/10.3989/scimar.03411.07C

Mente E, Gannon AT, Nikouli E, Kormas K (2016) Gut microbial communities associated with the molting stages of the giant freshwater prawn *Macrobrachium rosenbergii* prawn. Aquaculture 463:181–188

Mente E, Pierce GJ, Antonopoulou E, Stead D, Martin SAM (2017) Postprandial hepatic protein expression in trout *Oncorhynchus mykiss* a proteomics examination. Biochem Biophys Rep 9:79–85

Merchie G, Kontara E, Lavens P, Robles R, Kurmaly K, Sorgeloos P (1998) Effect of vitamin C and astaxanthin on stress and disease resistance of postlarval tiger shrimp, *Penaeus monodon* (Fabricius). Aquac Res 29(8):579–585

Meziti A, Ramette A, Mente E, Kormas KA (2010) Temporal shifts of the Norway lobster (*Nephrops norvegicus*) gut bacterial communities. FEMS Microbiol Ecol 74:472–484

Meziti A, Mente E, Kormas AK (2012) Gut bacteria associated with different diets in reared *Nephrops norvegicus*. Syst Appl Microbiol 35:473–482

Mie A, Andersen HR, Gunnarsson S, Kahl J, Kesse-Guyot E, Rembialkowska E, Quaglio G, Grandjean P (2017) Human health implications of organic food and organic agriculture: a comprehensive review. Environ Health 16:111. https://doi.org/10.1186/s12940-017-0315-4

Miller MR, Nichols PD, Bridle AR, Carter CG (2008) Increased elongase and desaturase gene expression with stearidonic acid enriched diet did not enhance long-chain omega 3 content of seawater Atlantic salmon (*Salmo salar* L.). J Nutr 138:2179–2185

Mitra G, Mukhopadhyay PK, Chattopadhyay DN (2005) Nutrition and feeding in freshwater prawn (*Macrobrachium rosenbergii*) farming. Aquafeed Formulation Beyond 2:17–19

Montero D, Izquierdo M (2010) Welfare and health of fish fed vegetable oils as alternative lipid sources to fish oil. In: Turchini GM, Ng WK, Tocher RD (eds) Fish oil replacement and alternative lipid sources in aquaculture feeds. CRC Press, Taylor and Francis, Boca Raton, pp 439–486

Montero D, Mathlouthi F, Tort L, Afonso JM, Torrecillas S, Fernandez Vaquero A, Negrin D, Izquierdo MS (2010) Replacement of dietary fish oil by vegetable oils affects humoral immunity and expression of proinflammatory cytokines genes in gilthead sea bream *Sparus aurata*. Fish Shellfish Immunol 29(6):1073–1081

Morales GA, Márquez L, de Rodrigañez MS, Bermúdez L, Robles R, Moyano FJ (2013) Effect of phytase supplementation of a plant-based diet on phosphorus and nitrogen bioavailability in sea bream *Sparus aurata*. Aquac Nutr. https://doi.org/10.1111/anu.12063

Mourente G (1996) In vitro metabolism of 14C-polyunsaturated fatty acids in midgut gland and ovary cells from *Penaeus kerathurus* at the beginning of sexual maturation. Comp Biochem Physiol B: Biochem Mol Biol 115:255–266. https://doi.org/10.1016/0305-0491(96)00111-3

Mourente G, Good JE, Bell JG (2005) Partial substitution of fish oil with rapeseed, linseed and olive oils in diets for European sea bass (*Dicentrarchus labrax* L.): effects on flesh fatty acid composition, plasma prostaglandins E2 and F2α, immune function and effectiveness of a fish oil finishing diet. Aquac Nutr 11:25–40

Moya-Falcon C, Thomassen M, Jakobsen JV, Ruyter B (2005) Effects of dietary supplementation of rapeseed oil on metabolism of [1-C-14] 18: 1 n-9, [1-C-14] 20: 3n-6, and [1-C-14]20: 4n-3 in Atlantic salmon hepatocytes. Lipids 40:709–717

Nandeesha MC, Gangadhara B, Varghese TJ, Keshavanath P (2000) Growth response and flesh quality of common carp, *Cyprinus carpio* fed with high levels of nondefatted silkworm pupae. Asian Fish Sci 13:235–242

Negas I, Alexis MN (1995) Partial substitution of fishmeal with soybean meal products and derivatives in diets for the gilthead sea bream (*Sparus aurata L.*). Aquac Res 27:147–156

Newton GL, Booram CV, Barker RW, Hale OM (1977) Dried *Hermetia illucens* larvae meal as a supplement for swine. J Anim Sci 44:395–400

Ng WK, Liew FL, Ang LP, Wong KW (2001) Potential of mealworm (*Tenebrio molitor*) as an alternative protein source in practical diets for African catfish, *Clarias gariepinus*. Aquac Res 32:273–280

Niu J, Tian L-X, Liu Y-J, Yang H-J, Ye C-X, Gao W, Mai K-S (2009) Effect of dietary astaxanthin on growth, survival, and stress tolerance of postlarval shrimp, *Litopenaeus vannamei*. J World Aquacult Soc 40(6):795–802

NRC (1993) Nutrient Requirements of Fish. Committee on Animal Nutrition. Board on Agriculture. National Research Council. National Academy Press, Washington, DC

NRC (2011) Nutrient Requirements of Fish and Shrimp. Committee on Nutrient Requirements of Fish and Shrimp. Board on Agriculture and Natural Resources. Division on Earth and Life Studies. National Research Council of the National Academies. The National Academies Press, Washington, DC. 376 pp

Ogunji JO, Kloas W, Wirth M, Schulz C, Rennert B (2006) Housefly maggot meal (magmeal): an emerging substitute of fishmeal in tilapia diets. Deutscher Tropentag 2006, Conference on International Agricultural Research for Development, Stuttgart-Hohenheim, Germany, 11–13 October 2006, 7 pp

Oliva-Teles A (2000) Recent advances in European sea bass and gilthead sea bream nutrition. Aquac Int 8:477–492

Oliva-Teles A, Gonçalves P (2001) Partial replacement of fishmeal by brewers yeast (Saccharomyces cerevisiae) in diets for sea bass (*Dicentrarchus labrax*) juveniles. Aquaculture 202:269–278

Øverland M, Tauson A-H, Shearer K, Skrede A (2010) Evaluation of methane-utilising bacteria products as feed ingredients for monogastric animals. Arch Anim Nutr 64:171–189

Øverland M, Karlsson A, Mydland LT, Romarheim OH, Skrede A (2013) Evaluation of *Candida utilis*, Kluyveromyces marxianus and *Saccharomyces cerevisiae* yeasts as protein sources in diets for Atlantic salmon (*Salmo salar*). Aquaculture 402–403:1–7

Özşahinoğlu I, Eroldoğan T, Mumoğullarında P, Dikel S, Engin K, Yılmaz AH, Arslan M, Sirkecioğlu AN (2013) Partial replacement of fish oil with vegetable oils in diets for european seabass (*Dicentrarchus labrax*): effects on growth performance and fatty acids profile. Turk J Fish Aquat Sci 13:819–825

Paibulkichakul C, Piyatiratitivorakul S, Kittakoop P, Viyakarn V, Fast AW, Menasveta P (1998) Optimal dietary levels of lecithin and cholesterol for blak tiger prawn *P. monodon* larvae and postlarvae. Aquaculture 167(1998):273–281

Palmegiano GB, Gai F, Gasco L, Lembo G, Spedicato MT, Trotta P, Zoccarato I (2009a) Partial replacement of fish meal by T-Iso in gilthead sea bream (*Sparus aurata*) juveniles diets. Ital J Anim Sci 8:869–871

Palmegiano GB, Gai F, Gasco L, Lembo G, Spedicato MT, Trotta P, Zoccarato I (2009b) Partial replacement of fish meal by T-Iso in gilthead sea bream (*Sparus aurata*) juveniles diets. Italian Journal of Animal Science 8: Proceedings of the 18th ASPA Congress, Palermo, June 9–12, 2009

Paripatananont T, Boonyaratpalin M, Pengseng P, Chotipuntu P (2001) Substitution of soy protein concentrate for fishmeal in diets of tiger shrimp *Penaeus monodon*. Aquac Res 32:369–374

Pereira TG, Oliva-Teles A (2002) Preliminary evaluation of pea seed meal in diets for gilthead sea bream (*Sparus aurata*) juveniles. Aquac Res 33:1183–1189

Pereira TG, Oliva-Teles A (2003) Evaluation of corn gluten meal as a protein source in diets for gilthead sea bream (*Sparus aurata* L.) juveniles. Aquac Res 34:1111–1117

Perera WMK, Carter CG, Houlihan DF (1995) The effect of replacing fishmeal with bacterial single cell protein on the nitrogen balance of rainbow trout (*Oncorhynchus mykiss* (Walbaum)). Br J Nutr 73:591–603

Peres H, Oliva-Teles A (1999) Influence of temperature on protein utilization in juvenile European seabass (*Dicentrarchus labrax*). Aquaculture 170:337–348

Peres H, Oliva-Teles A (2009) The optimum dietary essential amino acid profile for gilthead seabream (*Sparus aurata*) juveniles. Aquaculture 296:81–86

Piccolo G, Iaconisi V, Marono S, Gasco L, Loponte R, Nizza S, Bovera F, Parisi G (2017) Effect of *Tenebrio molitor* larvae meal on growth performance, in vivo nutrients digestibility, somatic and marketable indexes of gilthead sea bream (*Sparus aurata*). Anim Feed Sci Technol 226:12–20

Piedad-Pascual F (1984) Status of prawn (*Penaeus monodon*) feed development in the Philippines. In: Prawn Industry Development in the Philippines: Proceedings of the National Prawn Industry Development Workshop, 10–13 April 1984, Iloilo City, Philippines. Southeast Asian Fisheries Development Center, Aquaculture Department, Tigbauan, pp 75–82

Pratoomyot J, Bendiksen EA, Bell JG, Tocher DR (2010) Effects of increasing replacement of dietary fishmeal with plant protein sources on growth performance and body lipid composition of Atlantic salmon (*Salmo salar* L.). Aquaculture 305:124–132

Rajoka MI, Khan SH, Jabbar MA, Awan MS, Hashmi AS (2006) Kinetics of batch single cell protein production from rice polishings with *Candida utilis* in continuously aerated tank reactors. Bioresour Technol 97:1934–1941

Reed L, D'Abramo LR (1989) A standard reference diet for crustacean nutrition research. III. Effects on weight gain and amino acid composition of whole body and tail muscle of juvenile prawns *Macrobrachium rosenbergii*. J World Aquacult Soc 20:107–113

Rembiałkowska E (2007) Review quality of plant products from organic agriculture. J Sci Food Agric 87:2757–2762

Ringø E, Zhou Z, Vecino JLG, Wadsworth S, Romero J, Kroghdal A, Olsen RE, Dimitoglou A et al (2016) Effect of dietary components on the gut microbiota of aquatic animals: the never-ending story? Aquac Nutr 22:219–282

Rodrigues PM, Silva TS, Dias J, Jessen F (2012) Proteomics in aquaculture: applications and trends. J Proteome 3:4325–4345

Rodrigues PM, Martin SAM, Silva TS, Boonanuntanasarn S, Schrama D, Moreira M, Raposo C (2018) Proteomics in fish and aquaculture research. Chapter. In: de Almeida AM et al (eds) Proteomics in domestic animals: from farm to systems biology. Springer, Cham. https://doi.org/10.1007/978-3-319-69682-9_16

Romarheim OH, Øverland M, Mydland LT, Skrede A, Landsverk T (2011) Bacteria grown on natural gas prevent soybean meal-induced enteritis in Atlantic salmon. J Nutr 141:124–130

Rumbos OH, Karapanagiotidis IT, Mente E, Athanassiou CG (2018) The lesser mealworm Alphitobius diaperinus: a noxious pest or a promising nutrient source? Rev Aquac. https://doi.org/10.1111/raq.12300

Rungrassamee W, Klanchui A, Maibunkaew S, Chaiyapechara S, Jiravanichpaisal P, Karoonuthaisiri N (2014) Characterization of intestinal bacteria in wild and domesticated adult black tiger shrimp (*Penaeus monodon*). PLoS One 9(3):e91853. https://doi.org/10.1371/journal.pone.0091853

Ruyter B, Thomassen M (1999) Metabolism of n-3 and n-6 fatty acids in Atlantic salmon liver: stimulation by essential fatty acid deficiency. Lipids 34:1167–1176

Ruyter B, Røsjø C, Måsøval K, Einen O, Thomassen MS (2000a) Influence of dietary n-3 fatty acids on the desaturation and elongation of [1-^{14}C] 18:2 n-6 and [1-^{14}C] 18:3 n-3 in Atlantic salmon hepatocytes. Fish Physiol Biochem 23:151–158

Ruyter B, Røsjø C, Einen O, Thomassen MS (2000b) Essential fatty acids in Atlantic Salmon: effects of increasing dietary doses of n-6 and n-3 fatty acids on growth, survival and fatty acid composition of liver, blood and carcass. Aquac Nutr 6:119–127

Ruyter B, Røsjø C, Einen O, Thomassen MS (2000c) Essential fatty acids in Atlantic Salmon: time course of changes in fatty acid composition of liver, blood and carcass induced by a diet deficient in n-3 and n-6 polyunsaturated fatty acids. Aquac Nutr 6:109–117

Ruyter B, Røsjø C, Grisdale-Helland B, Rosenlund G, Obach A, Thomassen MS (2003) Influence of temperature and high dietary linoleic acid content on the elongation and desaturation of polyunsaturated fatty acids in Atlantic salmon hepatocytes. Lipids 38:833–839

Ryckebosch E, Bruneel C, Termote-Verhalle R, Goiris K, Muylaert K, Foubert I (2014) Nutritional evaluation of microalgae oils rich in omega-3 long chain polyunsaturated fatty acids as an alternative for fish oil. Food Chem 160:393–400

Sahu NP (2004) Nutrient requirement of Indian fishes of commercial importance: present status and future need. Indian Soc Fish Prof Newsl 'Indofish' VI(1–2):7–15

Santigosa E, Garcia-Meilan I, Maria Valentin J, Navarro I, Perez-Sanchez J, Angeles Gallardo M (2011a) Plant oils' inclusion in high fish meal-substituted diets: effect on digestion and nutrient absorption in gilthead sea bream (*Sparus aurata* L.). Aquac Res 42:962–974

Santigosa E, Garcia-Meilan I, Valentin JM, Perez-Sanchez J, Medale F, Kaushik S, Gallardo MA (2011b) Modifications of intestinal nutrient absorption in response to dietary fish meal replacement by plant protein sources in sea bream (*Sparus aurata*) and rainbow trout (Onchorynchus mykiss). Aquaculture 317:146–154

Sargent JR, Tocher DR, Bell JG (2002) The lipids. In: Halver, Hardy (eds) Fish nutrition. Academic Press, Elsevier Science, San Diego, pp 181–257

Sealey WM, Gaylord TG, Barrows FT, Tomberlin JK, McGuire MA, Ross C, St-Hilaire S (2011) Sensory analysis of rainbow trout, *Oncorhynchus mykiss*, fed enriched black soldier fly prepupae, *Hermetia illucens*. J World Aquacult Soc 42:34–45. https://doi.org/10.1111/j.1749-7345.2010.00441.x

Sitjà-Bobadilla A, Peña-Llopis S, Gómez-Requenia P, Médaleb F, Kaushikb S, Pérez-Sánchez J (2005) Effect of fish meal replacement by plant protein sources on non-specific defence mechanisms and oxidative stress in gilthead sea bream (*Sparus aurata*). Aquaculture 249:387–400

Skrede A, Mydland LT, Ahlstrom O, Reitan KI, Gislerod HR (2011) Evaluation of microalgae as sources of digestible nutrients for monogastric animals. J Anim Feed Sci 20:131–142

Smith DM, Tabrett SJ, Braclay MC (2001) Cholesterol requirement of subadult black tiger shrimp *Penaeus monodon* (Fabricius). Aquac Res 32(Suppl. 1):399–405

Soil Association (2009) Organic market report 2009. Retrieved on 15/07/2009 from http://www.soilassociation.org/Businesses/Marketinformation/tabid/116/Default.aspx

Sørensen M, Berge GM, Thomassen M, Ruyter B, Hatlen B, Ytrestøyl T, Aas TS, Åsgård T (2011) Today's and tomorrow's feed ingredients in Norwegian aquaculture. Nofima Report 52/2011, 68 pp

Sprague MJR, Dick JR, Tocher DR (2016) Impact of sustainable feeds on omega-3 long-chain fatty acid levels in farmed Atlantic salmon, 2006–2015. Sci Rep 6:21892. https://doi.org/10.1038/srep21892

St-Hilaire S, Cranfill K, McGuire MA, Mosley EE, Tomberlin JK, Newton L, Sealey W, Sheppard C, Irving S (2007a) Fish offal recycling by the black soldier fly produces a foodstuff high in Omega-3 fatty acids. J World Aquacult Soc 38:309–313

St-Hilaire S, Shepard C, Tomberlin JK, Irving S, Newton L, McGuire MA, Mosley EE, Hardy RW, Sealey W (2007b) Fly prepupae as a feedstuff for rainbow trout, *Oncorhynchus mykiss*. J World Aquacult Soc 38:59–67

Tacon AGJ (2002) Thematic review of feeds and feed management practices in shrimp aquaculture. Report prepared under the World Bank, NACA, WWF and FAO

Tacon AGJ, Metian M (2015) Feed matters: satisfying the feed demand of aquaculture. Rev Fish Sci Aquac 23(1):1–10. https://doi.org/10.1080/23308249.2014.987209

Tarnecki AM, Burgos FA, Ray CL, Arias CR (2017) Fish intestinal microbiome: diversity and symbiosis unravelled by metagenomics. J Appl Microbiol. https://doi.org/10.1111/jam.13415

Thebault H, Alliot E, Pastoureaud A (1985) Quantitative methionine requirement of juvenile seabass (*Dicentrarchus labrax*). Aquaculture 50:75–87

Thomassen MS, Røsjø C (1989) Different fats in feed for salmon. Influence on sensory parameters, growth rate and fatty acids in muscle and heart. Aquaculture 79:129–135

Tibaldi E, Kaushik SJ (2005) Amino acid requirements of Mediterranean fish species. In: Montero D, Basurco B, Nengas I, Alexis M, Izquierdo M (eds) Mediterranean fish nutrition, Cahiers Options Méditerranéennes; n. 63. CIHEAM, 2005, Zaragoza, pp 59–65

Tiwari JB, Sahu NP (1999) Possible use of soyaphospholipid as a source of lipid in the post larval diet of *Macrobrachium rosenbergii*. J Aquac Tropics 14:37–46

Tocher, DR; Bell, JG; Dick, JR, Henderson, RJ, McGhee, F., Michell, D., Morris, PC., 2000. Polyunsaturated fatty acid metabolism in Atlantic salmon (*Salmo salar*) undergoing parr-smolt transformation and the effects of dietary linseed and rapeseed oils. Fish Physiol Biochem 23(1) 59–73.

Torrissen OJ, Christiansen R (1995) Requirements for carotenoids in fish. Appl Ichtyol 11:225–230

Torstensen BE, Bell JG, Rosenlund G, Henderson RJ, Graff IE, Tocher DR, Lie Ø, Sargent JR (2005) Tailoring of Atlantic salmon (*Salmo salar* L) flesh lipid composition and sensory quality by replacing fish oil with a vegetable oil blend. J Agric Food Chem 53:10166–10178

Torstensen BE, Espe M, Sanden M, Stubhaug I, Waagbø R, Hemre GI, Fontanillas R, Nordgarden U, Hevrøy EM, Olsvik P, Berntssen MHG (2008) Novel production of Atlantic salmon (*Salmo salar*) protein based on combined replacement of fish meal and fish oil with plant meal and vegetable oil blends. Aquaculture 285:193–200

Torstensen BE, Ruyter B, Sissener N, Østbye TK, Waagbø R, Jørgensen SM, Ytteborg E, Rud I, Liland N, Mørkøre T, Dessen, JE (2013) "Fett for fiskehelse" Utredning: Effekter av endret fettsyresammensetning i fôr til laks relatert til fiskens helse, velferdog robusthet. Nofima/NIFES/FHF. http://www.nofima.no/filearchive/900889-sluttrapport-fett-for-fiskehelse-godkjent-28062013.pdf

Tulli F, Zittelli GC, Giorgi G, Poli BM, Tibaldi E, Tredici MR (2012) Effect of the inclusion of dried *Tetraselmis suecica* on growth, feed utilization, and fillet composition of European Sea Bass juveniles fed organic diets. J Aquat Food Prod Technol 21:188–197

Van Huis A, Oonincx DGAB (2017) The environmental sustainability of insects as food and feed. A review. Agron Sustain Dev 37:43. https://doi.org/10.1007/s13593-017-0452-8

Velasco M, Lawrence AL, Castille FL, Obaldo LG (2000) Dietary protein requirement for *Litopenaeus vannamei*. In: Cruz-Suárez LE, Ricque-Marie D, Tapia-Salazar M, Olvera-Novoa MA, Civera-Cerecedo R (eds) Avances en Nutrición Acuícola V. Memorias del V Simposium Internacional de Nutrición Acuícola. 19–22 Noviembre, 2000. Mérida

Weir G (1998) The use of proteins in aquaculture feeds. Biotechnology in the feed industry (1998) 605–611. UCAAB L3260

Wilson RP (2002) Amino acids and proteins. In: Hardy RW, Halver J (eds) Fish nutrition. Academic, Amsterdam, pp 143–179

Wouters R, Lavens P, Nieto J, Sorgeloos P (2001) Penaeid shrimp broodstock nutrition: an updated review on research and development. Aquaculture 202:1–21

Yang Y, Wang Y, Lu Y, Li Q (2011) Effect of replacing fish meal with soybean meal on growth, feed utilization and nitrogen and phosphorus excretion on rainbow trout (*Oncorhynchus mykiss*). Aquac Int 19:405–419

Index

A
Accreditation bodies, 18, 20, 32
Active metabolic rate (AMR), 124
Algae, viii, 11, 15, 27, 36, 81, 86–89, 97, 98, 109, 133, 135, 142, 146, 159, 165, 169
Algal blooms, 94, 105, 122
α-linolenic (ALA, 18:3n-3), 80, 164
Amino acids, 7, 79, 84, 87, 99, 127, 144, 149, 153–157, 159–161, 173, 174
Antioxidants, 98, 99, 143
Aquafeed technology, x, 142
Artemia, 81–85, 87–90, 93, 96, 97, 99, 130
Astaxanthin, 29, 83, 150, 151, 169

B
Bacteria, 95, 107, 115, 129, 133, 150, 158, 159, 170
Best linear unbiased prediction (BLUP) methodology, 67
Biofouling, 111
Biological filtration, 106
Biosecurity, x, 30, 69, 87, 134–135, 158
Breeding, ix, 3, 5, 7, 14, 26, 28, 29, 44, 56, 65–73, 97, 133, 135
Brood stock management, 98

C
Cage systems, x, 110–112
Carbon dioxide, 121, 123, 128
Carbon footprint, 104, 110
Carnivorous species, 5, 15, 142, 153, 154, 158
Chemometrics, 73
Chronic stress, x, 121, 122
Cleaning and disinfection products, ix, 26
Codex Alimentarius, 9
Communication, x, 19, 32, 55, 59, 60, 162
Consumers, vii–x, 2, 3, 6–8, 11–13, 19, 20, 24, 31, 54–58, 60, 73, 112, 113, 141, 143, 159–161, 173
Consumers' attitudes, 47, 55–57
Control body, 21, 32, 33, 37
Copepods, 81, 83, 93, 165
Cost structures, 49
Crustaceans, viii–x, 15, 26, 28–30, 42, 86, 94–95, 143, 145, 150, 157–158, 160, 165–167, 169–171, 176

D
Delegated and Implementing Acts, 34
Density, 72, 80–82, 87, 91, 93, 111, 122, 123, 125, 126
Depreciation, 49
Diets, 2, 4, 15, 80, 82, 83, 85–87, 89, 93, 131, 144, 151–153, 155–157, 159–168, 170, 171, 174
Disease prevention, ix, 26, 30, 31, 33, 44, 113, 120, 131, 134
Diseases, ix, x, 6, 25, 30, 65–69, 73, 87, 94, 107, 108, 113, 115, 121, 125, 127, 129–131, 133–135, 155
Dissolved oxygen, 104, 105, 111, 121, 122
Distribution costs, 52, 59
Docosahexaenoic acid (DHA, 226n-3), 80, 164
Domestication, 12, 97, 98, 171

G. Lembo, E. Mente (eds.), *Organic Aquaculture*,
https://doi.org/10.1007/978-3-030-05603-2

E

Eggs, 79, 81, 83, 85, 86, 91, 93
Eicosapentaenoic acid (EPA, 20:5n-3), 80, 164
Emulsifiers, 96, 98, 99, 165
Enrichment, x, 81, 82, 88, 96, 97, 99, 107, 111, 120
Enzymes, 84, 145, 150, 156, 157, 162
EPA+DHA, 162, 166
Essential fatty acids (EFA), 4, 80, 81, 83, 88, 162
EU organic control system, ix, 31–32
EU organic regulations, 18, 23, 24, 36–38, 96, 171
European Council and the European Commission, 142
European Parliament, 29, 31, 133, 142, 144, 145, 160
European regulation, ix, 20, 133, 142
Expressed sequence tag (EST), 72

F

Fatty acids, 4, 80, 81, 83, 85, 86, 88, 96, 98, 99, 125, 153, 155, 161–166, 173
Feed additives, 130, 133, 143, 144, 149
Feed composition, x, 142
Feed efficiency, x, 68, 72, 73, 95
Feeding fish and crustaceans, ix, 26, 176
Feed ingredients, x, 84, 142–171, 173, 176
Feed loading, 105
Feeds, 2, 25, 42, 68, 80, 105, 120, 142
Fermentation, 7, 29, 99, 144, 149, 150, 156, 159, 165, 170, 174
Fish, 4, 25, 41, 67, 80, 105, 119, 143
Fishmeal, 5, 29, 142–146, 151–161, 167, 172, 173, 176
Fish oil, 15, 29, 30, 98, 99, 142–146, 151, 153, 156–158, 161–166, 172–174, 176
Fixed costs, 47, 48
Flow-through systems, 56, 105–106, 108
Food intake, 121, 142
Formulation, x, 7, 84, 85, 87–89, 98, 111, 142, 144, 151, 159, 170, 173, 174, 176
Freshwater, 8, 42, 70, 79, 85–86, 91, 95, 98, 104, 106, 107, 112, 122, 128, 131, 142, 145, 146, 162, 163, 168

G

Genetically modified organisms (GMO), 26, 143, 151
Genetic gains, ix, 65, 66
Genomics, 65–73, 175
Genomic-wide association studies (GWAS), 71
Genotype by environment interaction (GxE), ix, 69–71, 73
Genotype re-ranking, 70
Guarantee and certification systems, viii
Gut microorganisms, 171

H

Hatcheries, ix, x, 28–30, 73, 81, 83, 86–92, 94–95, 104, 111, 112, 134, 143, 145
Hatching, 80, 81, 91, 93–96
Herbal remedies, 130
Heterogeneity of genetic variance, 70
Heterozygosity-fitness correlation, 71
High performance clusters, 71
Histidine, 29, 73, 99, 144, 149, 160
Holistic conception, 3
Housing conditions, 120
HUFAs, 88, 161, 163–165
Husbandry practices, ix, 25, 26, 28, 29, 44, 120

I

IFOAM Organics International, 1–3, 6, 9
Immunostimulants, 30, 129–131, 134
Import system, 37, 38
Inbreeding management, 67
Insect meal, 160
Insects, 5, 7, 16, 158, 160
Inspection visits, 34
Integrated multi-trophic aquaculture (IMTA), x, 105, 112–116
Investments, 6, 48, 52, 110, 115
ISO/IEC 17011:2017, 18

L

Labelling, ix, 2, 8, 9, 12, 16, 17, 19, 20, 23, 24, 26, 31, 55, 57, 60
Labour costs, 49
Larvae, ix, 73, 79, 81–87, 90–98, 158, 160, 161
Larval development, 79–83, 86, 87, 96
Life cycles, 14, 36, 83, 94
α-Linoleic (LA, 18:2n-6), 80, 164
Load of nutrients, 105

M

Maintenance, 3, 4, 12, 16, 49, 103, 105, 125, 134
Marine, 4, 42, 67, 79, 103, 122, 143
Marine species, 81, 91, 153–157, 161, 165–168

Marker assisted selection (MAS) programs, 72, 73
Market size, 52, 59
Mesocosm, 92–94, 96
Metabolite profiling, 73
Metabolomics, 71, 73
Microalgae, 143, 159, 169
Microbiota, 131, 132, 170, 171, 174, 176
Microencapsulated diets, 81, 89
Minerals, 27, 30, 96, 143, 144, 148, 153, 155, 166–167
Molluscs production, viii, ix, 15, 26, 145
Morphological abnormalities, 121
Mysis, 87, 88

N
Native species, 56, 113
Nauplii, 81–88, 94, 96, 97
Next generation sequencing (NGS), 71, 72
Nitrogen, 113, 155, 156
Nutrient requirements, 84, 155, 176
Nutrition, vii, viii, x, 3, 4, 15, 16, 44, 72, 79, 86–90, 109, 131, 141–176
Nutritional physiology, x, 142, 153, 170, 175
Nutritional requirements, 79, 142–145, 151

O
Oils, 15, 29, 79, 127, 142
Omics, 71–73
Operational profit, 52
Organic fertilizers, 97, 98, 172
Organic market, 8, 12, 13
Organic matter, 25, 107, 108, 110, 112, 114, 135
Organic operators, 5, 12, 32–34, 37
Organic principles, viii, x, 2–7, 11, 109, 143, 146, 155, 166, 174
Organic standards, vii, viii, 2, 4, 5, 7, 10–12, 17, 44, 47, 55, 56, 96, 99, 109, 151–153, 176
Organic trimmings, 15, 29, 143, 145, 157, 174
Origin of the aquaculture animals, ix, 28
Owner remuneration, 52
Oxygenation, 95, 107, 126

P
Participatory Guarantee Systems (PGS), 19
Pedigree, 65–67
Phospholipids, 83, 88, 153, 165, 166
Phosphorus, 113, 156, 157, 160, 166, 167
Physiological and behavioural needs, 120, 142, 175
Phytase, 156
Phytoplankton, viii, x, 30, 91, 93, 96, 98, 143, 145, 146
Pigments, 80, 86, 95, 96, 99, 151, 169
Plant proteins, 98, 148, 154–156, 158
Pollution, 5, 14, 29, 105, 115, 129
Pond systems, x, 109–110
Prey, x, 5, 80, 81, 83–85, 87, 95, 96, 99, 157
Price premiums, 50, 57–59
Principle of care, 7
Principle of ecology, 4, 5
Principle of fairness, 6, 7
Principle of health, 3, 4
Probiotics, 30, 84, 131–133, 174
Processed animal protein (PAP), 160
Production costs, ix, 48–50, 81, 85, 93
Product quality, x, 3, 16, 30, 73, 142, 144
Profitability, ix, 50–52, 59
Protein, 2, 41, 71, 84, 110, 126, 143
Proteomics, 71, 72

Q
Quantitative trait locus (QTL), 66, 72

R
Recirculating aquaculture systems (RAS), 5, 105, 107–109, 121
Recycling, x, 4, 108, 112, 159
Renewable energy, 6, 108
Reproduction, viii, 67, 72, 79, 97, 124, 158, 162, 163, 167, 169
Resistance, 2, 25, 65, 66, 68, 69, 73, 80, 89, 90, 123, 129, 131, 132
Restriction site-associated DNA sequencing (RAD-seq), 71
Reuse of water, 106, 109
Risk evaluation, 34
Risk of escape, 28, 108
RNA-sequence (RNA-seq), 72
Rotifers, 28, 81–83, 85, 87, 93, 96, 99, 145

S
Salmon, 11, 45, 66, 99, 111, 122, 143
Salmonids, 4, 29, 67, 85, 111, 125, 135, 144, 149, 153–157, 159, 161–170, 176
Scope for activity (SFA), 124, 125, 164
Sea bass, 45, 46, 49–51, 53, 69, 70, 82, 93, 94, 122, 124, 125, 127, 128, 153–157, 159–168

Sea bream, 45, 46, 49–51, 53, 67, 69, 70, 122, 124, 125, 127, 128, 153–157, 159–168, 171
Seaweed production, 24, 27
Seaweeds, viii, 24, 26, 30, 109, 112–115, 131, 143, 146, 148
Selective breeding, ix, 65–73
Sexual dimorphism, 69
Shot-gun sequencing, 71
Shrimps, 14, 29, 50, 66, 81, 113, 130, 142
Single nucleotide polymorphisms (SNPs), 71
Skewed sex ratio stocks, 69
Slaughtering, 29, 126–129
Social inclusion, 46
Stakeholders' awareness, 12
Standard metabolic rate (SMR), 124
Stocking densities, x, 6, 16, 28, 30, 44, 47, 49, 50, 93, 94, 105, 106, 108, 109, 120, 123–126, 174
Stress coping styles (SCS), 119

T
The gastrointestinal tract microbiota (GIT), 170, 171
Trade opportunities, 46
Traits, 65–73, 119, 154, 160, 171
Transcriptomics, 71, 72
Trout, vii, 45, 46, 49, 50, 52, 53, 57, 66, 69, 70, 95, 110, 122, 124, 128, 130, 132, 150, 159, 161, 162, 164, 170

U
Un-ionized ammonia-nitrogen, 121
US Department of Agriculture's National Organic Program (USDA NOP), 9, 10
UV radiation, 107, 109

V
Vaccinations, 30, 129, 134
Veterinary treatments, ix, x, 16, 26, 30, 31, 33, 44, 119–135
Vitamins, 83, 96, 98, 99, 143, 149, 153, 155, 166–169

W
Waste products, 105, 110, 144, 152
Water quality, x, 6, 7, 14, 28, 81, 87, 104, 105, 108, 109, 111, 120–123, 127, 134
Weaning, 79–99
Welfare indicators, 121

Y
Yeasts, 81, 98, 132, 133, 150, 159, 165, 169
Yolk, 79, 86, 91, 95

Z
Zoea, 86, 88, 90
Zooplankton, x, 28, 30, 80, 83, 87, 91, 93, 94, 96, 98, 145, 159, 161, 165

MIX
Papier aus verantwortungsvollen Quellen
Paper from responsible sources
FSC® C105338

FSC
www.fsc.org

If you have any concerns about our products,
you can contact us on
ProductSafety@springernature.com

In case Publisher is established outside the EU,
the EU authorized representative is:
Springer Nature Customer Service Center GmbH
Europaplatz 3, 69115 Heidelberg, Germany

Printed by Libri Plureos GmbH
in Hamburg, Germany